Biotechnology: Pharmaceutical Aspects

For other titles published in the series, go to
www.springer.com/7364

Biotechnology: Pharmaceutical Aspects

Volume I: *Pharmaceutical Profiling in Drug Discovery for Lead Selection*
R.T. Borchardt, E.H. Kerns, C.A. Lipinski, D.R. Thakker, B. Wang

Volume II: *Lypophilization of Biopharmaceuticals*
H.R. Constantino, M.J. Pikal

Volume III: *Methods for Structural Analysis of Protein Pharmaceuticals*
W. Jiskoot, D.J.A. Crommelin

Volume IV: Optimizing *the "Drug-Like" Properties of Leads in Drug Discovery*
R.T. Borchardt, E.H. Kerns, M.J. Hageman, D.R. Thakker, J.L. Stevens

Volume V: *Prodrugs: Challenges and Rewards, Parts 1 and 2*
V.J. Stella, R.T. Borchardt, M.J. Hageman, R. Oliyai, H. Maag, J.W. Tilley

Volume VI: *Solvent Systems and Their Selection in Pharmaceutics and Biopharmaceutics*
P. Augustijns, M.E. Brewster

Volume VII: *Drug Absorption Studies: In Situ, In Vitro and In Silico Models*
C. Ehrhardt, K.J. Kim

Volume VIII: *Immunogenicity of Biopharmaceuticals*
M. van de Weert, E. H. Møller

Volume IX: *Advances in Bioactivation Research*
A. Elfarra

Volume X: *Nanotechnology in Drug Delivery*
M. M. de Villiers, P. Aramwit, G. S. Kwon

Volume XI: *Current Trends in Monoclonal Antibody Development and Manufacturing*
S.J. Shire, W. Gombotz, K. Bechtold-Peters, J. Andya

Volume XII: *Pharmaceutical Stability Testing to Support Global Markets*
K. Huynh-Ba

Pharmaceutical Stability Testing to Support Global Markets

Edited by

Kim Huynh-Ba

Pharmalytik
Newark, DE
USA

Springer

AAPS PRESS

Editor
Kim Huynh-Ba
Pharmalytik
Newark, DE
USA
kim.huynhba@pharmalytik.com

ISBN 978-1-4419-0888-9 e-ISBN 978-1-4419-0889-6
DOI 10.1007/978-1-4419-0889-6
Springer New York Dordrecht Heidelberg London

Library of Congress Control Number: 2009939690

Printed on acid-free paper

Springer is part of Springer Science+Business Media (www.springer.com)

Preface

The International Conference of Harmonization (ICH) has worked on harmonizing the stability regulations in the US, Europe, and Japan since the early 1990s. Even though the Stability Guidelines Q1A (R2) was issued over a decade ago, issues surrounding this arena continue to surface as the principles described in the guideline are applied to different technical concentrations. As a result, the stability community has continued to discuss concerns and find ways of harmonizing regulatory requirements, streamlining practices, improving processes in order to bring safe and effective medical supplies to the patients around the world.

In 2007, the American Association of Pharmaceutical Scientists (AAPS) Stability Focus Group organized two workshops – the Stability Workshop and the Degradation Mechanism Workshop. These meetings attracted many industry scientists as well as representatives from several regulatory agencies in the world to discuss important topics related to pharmaceutical stability practices.

Recognizing the importance of documenting these discussions and with the permission of AAPS, I have worked with speakers to assemble a collection of 30 articles from presentations given at these two meetings, mainly the Stability Workshop. I trust that this book will be beneficial to all of you in providing guidance and up-to-date information for building quality stability programs.

Freedom of our mind is
Mother of all inventions.

Thai Huynh-Ba

Editorial Notes

The Stability Workshop was held on 10–12 September 2007, organized by the American Association of Pharmaceutical Scientists (AAPS) Stability Focus Group. It was a 3-day event with multiple parallel tracks of technical discussions. The transcript of the workshop was available as a service from the AAPS.

Contributing authors are responsible for the content and ideas expressed in their articles and presentations. Great efforts were made to assure that the articles included in this book are as accurate and current as possible; however, there might be occasional errors. The editor wishes to hold no responsibility for, nor does she endorse, the material published in this publication made by presenters at this meeting.

About the Editor

Kim Huynh-Ba is the founder and Technical Director of Pharmalytik (http://www.pharmalytik.com). She has 22 years of experience in various analytical areas of pharmaceutical development and a primary focus in stability sciences. Prior to Pharmalytik, she held positions in drug development at Astra Zeneca (formerly ICI Americas), DuPont Merck, DuPont Pharmaceuticals, Bristol Myers Squibb, and Wyeth Vaccines. She has been advising pharmaceutical companies including companies operating under Consent Decree on harmonization and optimization of analytical best practices since 2001.

In addition to her consulting activities, Kim is a short course instructor and organizer on topics ranging from cGMP compliance and quality issues to stability programs under sponsorship of global organizations such as American Chemical Society (ACS), American Association of Pharmaceutical Scientists (AAPS), Pittsburgh Conference, many other international training groups. She is the founder and past co-chair of the AAPS Stability Focus Group, and an active member of the Pharmaceutical Stability Discussion Group (PSDG). She serves on the Governing Board of Eastern Analytical Symposium (EAS). She is currently chair of the AAPS APQ e-Learning Committee, and the 2008 EAS Short Course Program. She is a member of USP's Prescription/Non-Prescription Stakeholder Forum and also USP Reference Standard Project Team. She is a visiting professor at Rowan University and an adjunct professor in the Professional Science Master in Analytical Chemistry Program at Illinois Institute of Technology.

Kim Huynh-Ba is a recipient of the 2008 AAPS APQ Service Award and 2008 Recognition Award of AAPS Regulatory Section. She also received the 2001 DPCAA Leadership Award.

Kim Huynh-Ba has authored numerous technical publications and book chapters. She is invited frequently to present at national and international conferences. She is the editor of the *Handbook of Stability Testing in Pharmaceutical Development: Regulations, Methodologies and Best Practices*, which has been recognized as a practical reference book in the stability community.

Acknowledgments

The editor wishes to acknowledge with gratitude the substantial help and encouragement that she has received from the following individuals:

Special thanks are due to the workshop planning committee. This group provided tireless efforts to build one of the most comprehensive conferences of its time. They also served as moderators and speakers of the meetings.

Special thanks are due to all contributing authors. They exemplify the expertise of their field of interest. Many speakers had traveled long distance from other continents to Bethesda, MD, USA. They participated at the meetings with great enthusiasm and cooperation. Disregarding their demanding schedule, they have transformed their presentation to valuable manuscript for this volume.

Special thanks are due to the following individuals who have critically reviewed the papers in this book: Richard Adams, Anne-Françoise Aubry, Steve Baertschi, Ellen Carsch, Richard Currey, Mary Ann Gorko, Rita Hassall, Jianmei Kochling, Judy Lin, Karen Lucas, Betty Mellellus-Boutros, Mikie McGinnes, Alvin Melveger, Andrea Panaggio, James Shea, Paula Youngberg-Webb, Lawrence Yu, and Manuel Zahn. Also thanks are due to Ani Sarkahian for the final editing of all chapters.

Special thanks are due to members of several organizations, specifically the U.S. Food and Drug Administration (FDA), the World Health Organization (WHO), the United State Pharmacopeia (USP), the Eastern Analytical Symposium and Exposition (EAS), the Pharmaceutical Research and Manufacturers of America (PhRMA), the Generic Pharmaceutical Association (GPhA), and the Consumer Health Pharmaceutical Association (CHPA) for their advices of technical content and/or co-sponsorship.

Special thanks are due to AAPS staff members, Elizabeth Marburger and Megan Reese, who helped with the logistics of this workshop, and Andy Cohn, who has provided the transcripts for this meeting.

Special thanks to Pharmalytik for providing a special publication grant to complete this book volume, and Mr. Paul Huynh for his assistance in preparing materials for this volume.

And my deepest appreciation goes to my family, my husband Thai, my sons John and James, for their encouragement and my parents, Mr. Hong Nguon Cao and Mrs. Kimhoa Ngoc Bach, for their total support.

Contents

Section IV Stability Data and Operational Practices

Contributors

Robert Aitchison
Pfizer Global Research and Development, Kent, UK

Mark D. Argentine
Eli Lilly and Company, Indianapolis, IN, USA

Conny Axelsson
AstraZeneca Global Quality Operations, Södertälje, Sweden

Steven W. Baertschi
Eli Lilly and Company, Lilly Research Laboratories, Indianapolis, IN, USA

Michael D. Barron
Holly Springs, NC, USA

Jon V. Beaman
Pfizer Global Research and Development, Kent, UK

Gaurav Bedse
National Institute of Pharmaceutical Education and Research, Nagar, India

Juma Bridgewater
Vertex Pharmaceuticals, Inc., Cambridge, MA, USA

Gary Buehler
Food and Drug Administration, Center for Drug Evaluation
and Research, Office of Generic Drugs, Rockville, VA, USA

Frank Diana
Endo Pharmaceuticals, Chadds Ford, PA, USA

Abbie Gentry
McNeil Consumer Healthcare, Fort Washington, Pennsylvania, USA

Yushen Guo
Achaogen, Inc., South San Francisco, CA, USA

Abhay Gupta
U.S. Food and Drug Administration, Division of Product Quality Research, Silver Spring, MD, USA

Kim Huynh-Ba
Pharmalytik, Newark, DE, USA

Patrick J. Jansen
Eli Lilly and Company, Indianapolis, IN, USA

Mansoor A. Khan
U.S. Food and Drug Administration, Division of Product Quality Research, Center for Drug Evaluation and Research, Silver Spring, MD, USA

Jianmei Kochling
Vertex Pharmaceuticals, Inc., Cambridge, MA, USA

Sabine Kopp
World Health Organization, WHO Expert Committee on Specifications for Pharmaceutical Preparations, Quality Assurance of Medicines, Geneva, Switzerland

Vijay Kumar
National Institute of Pharmaceutical Education and Research, Nagar, India

Karen L. Lucas
Johnson & Johnson Consumer & Personal Products Worldwide, Morris Plains, NJ, USA

Jairaj (Jai) Mehta
JM Pharma, LLC, Peoria, AZ, USA

Ganapathy Mohan
Sanofi-aventis, Collegeville, PA, USA (when the presentation was conducted)
Current affiliation: Merck and Co., West Point, PA, USA

Justina A. Molzon
Food and Drug Administration, Center for Drug Evaluation and Research, International Programs, Rockville, VA, USA

Eda Ross Montgomery
Vertex Pharmaceuticals Co., Cambridge, MA, USA

Redouan Naji
Vertex Pharmaceuticals, Inc., Cambridge, MA, USA

Jeffrey T. Needels
Novartis Consumer Health, Lincoln, NE, USA

Earl Nordbrock
Nordbrock Consulting, Draper, UT, USA

Horacio N. Pappa
United Stated Pharmacopeia, Rockville, MD, USA

Amrit Paudel
National Institute of Pharmaceutical Education and Research, Nagar, India

Rachael Carlisle Roehrig
Consumer Healthcare Products Association, Washington, DC, USA

Karen A. Russo
United States Pharmacopeial, Rockville, MD, USA

Abdel Aziz Saleh
Previously World Health Organization, Eastern Mediterranean Regional Office, Cairo, Egypt

Vilayat A. Sayeed
U.S. Food and Drug Administration, Division of Chemistry III, Rockville, MD, USA

Mary W. Seibel
Procter and Gamble Health Care, Cincinnati, OH, USA

Rakhi B. Shah
U.S. Food and Drug Administration, Division of Product Quality Research, Center for Drug Evaluation and Research, Silver Spring, MD, USA

Saranjit Singh
National Institute of Pharmaceutical Education and Research, Nagar, India

Lucky S. Slamet
National Agency of Drug and Food Control (NADFC), Jakarta Pusat, Republic of Indonesia

Mark A. Staples
Cusp PharmaTech Consulting, Cambridge, MD, USA

Rhishikesh Thakare
National Institute of Pharmaceutical Education and Research, Nagar, India

Saji Thomas
Par Pharmaceuticals, Inc., New York, NY, USA

Paula J. Youngberg Webb
Baxter Healthcare Corporation, Round Lake, IL, USA

Duu-Gong Wu
PharmaNet, Inc., Princeton, NJ, USA

Section I

Chapter 1

Introduction

Kim Huynh-Ba

1. Introduction

The articles in this book are based on presentations made at the AAPS Workshop on "*Pharmaceutical Stability Testing to Support Global Markets*" in Bethesda, Maryland, September 10–12, 2007. This workshop brought together 45 U.S. and international speakers, representing all four different climatic zones. Its content drew over 350 attendees from around the globe. The authors discussed current trends in global stability testing and presented specific technical issues that still need to be resolved to fully harmonize stability practices.

2. Background on the Stability Focus Group and the Workshop

2.1. Stability Focus Group

The AAPS Stability Focus Group was formed in March 2006 under the umbrella of the Analysis and Pharmaceutical Quality (APQ) Section of AAPS. Currently, the focus group is affiliated to both the APQ and Regulatory Sciences (RS) Sections and is led by a Steering Committee established in April 2006. The goal of the Focus Group was to organize technical programs where current issues related to stability testing had to be discussed. The September 2007 workshop was the first in this series of technical meetings.

2.2. Background of the Workshop

Stability is a critical quality attribute of all pharmaceutical products, and therefore stability testing is a crucial component of drug development process. Companies rely on the stability data to establish the shelf life of marketed pharmaceutical products and ensure the efficacy and safety of drugs. Many guidelines have been written in this arena; however, new issues continue to raise and challenge the established practices.

From: *Pharmaceutical Stability Testing to Support Global Markets*: Pharmasp,
Edited by: K. Huynh-Ba, DOI 10.1007/978-1-4419-0889-6_1,

FDA issued its first stability guidance in 1987. There were considerable efforts in the early 1990s to harmonize the stability practices within the International Conference of Harmonization (ICH) regions (US, Europe and Japan). In September 1993, ICH published its draft Stability guidelines Q1A. Following this harmonization initiative, several other stability related guidelines were started. In 1998, the FDA published a draft stability guidance which combined ICH Q1AR2 with several other ICH guidelines. This document has been a foundation for much of the stability practices since then. In 2004, another milestone was achieved when ICH issued Q1F for recommendation of stability program conducted to support Zone 3 and 4. Subsequently, concerns were raised with the testing condition for extremely hot and humid climate by the Association of South East Asian Nations (ASEAN).

In June 2006, the FDA withdrew the 1987 Stability Guidance as well as the 1998 Stability Draft Guidance, and consequently, ICH withdrew the ICH Q1F guideline in July of the same year. Much of the discussion at the time was about the paradigm change at the FDA that introduced the Quality-by-Design (QbD) concepts in drug development.

The withdrawal of Stability Guidances in June 2006 has left a regulatory void for drug manufacturers in deciding what stability storage conditions to use in support of global submissions. In this workshop, we brought together multidisciplinary industry experts, as well as regulatory experts, for the different climatic zones to address the stability concerns from their point of view. In addition, several technical issues that impact global submissions were also discussed and provided a framework for developing an effective and science-based stability program. The workshop was comprised of several sessions organized in parallel tracks that each explored different stability related challenges, such as monitoring impurities, evaluating shipping excursions, setting specifications, estimating expiry, stability testing of Over-The-Counter (OTC), Nutraceuticals and Generics drug products. Discussion also addressed solid state physical stability of drug substance, which has significant effects on both chemical stability and bioavailability of the drug products, as well newer issues such as stability of split tablets, repackaged products, and genotoxic impurities. As some companies turn to Contract Research Organizations (CROs) for their analytical testing, the workshop also covered the technical and organizational challenges of managing drug development in a virtual environment.

This meeting was sponsored by the Eastern Analytical Symposium and Exposition (EAS), Pharmaceutical Research and Manufacturers of America (PhRMA), Generic Pharmaceutical Association (GPhA), and Consumer Healthcare Product Association (CHPA).

3. Meeting Planning Committee

This workshop was developed in order to provide the global regulatory perspectives on stability programs and technical issues of stability testing in twenty-first century. The content was designed by the following planning committee members:

Kim C. Huynh-Ba, M.Sc., Pharmalytik, *Co-chair, (EAS)*
Saji K. Thomas, M.Sc., Par Pharmaceutical, Inc., *Co-chair*
Mark S. Alasandro, Ph.D., Merck & Company, Inc (PhRMA).
Steven W. Baertschi, Ph.D., Eli Lilly and Company
Nicholas Cappuccino, Ph.D., Eagle Pharmaceuticals, Inc (GPhA).
Dilip R. Choudhury, Ph.D., Allergan Inc.
Abbie E. Gentry, Ph.D., McNeil Consumer Healthcare
Yushen Guo, Ph.D., Achaogen Inc.
Mansoor Khan, Ph.D., U.S. Food and Drug Administration
Karen L. Lucas, B.Sc., Johnson and Johnson Consumer Groups (CHPA)
Prabu Nambiar, Ph.D., Vertex Pharmaceuticals
Nanda K. Subbarao, Ph.D., Biologics Consulting Group
Thirunellai G. Venkateshwaran, Ph.D., Wyeth Research
Paula J. Youngberg Webb, M.Sc., Baxter Healthcare Company
Manuel Zahn, Ph.D., 3R Pharma Consulting
Melvin H. Weinswig, Ph.D., University of Wisconsin-Madison (Cont. Ed.)

4. Introduction of this Volume

This publication is coordinated and edited by Ms. Kim Huynh-Ba, from Pharmalytik, Newark, Delaware. This book contains three main sections:

The Stability Studies in a Global Environment Section introduces regulatory initiatives on global stability submission. Several models of harmonization efforts are explained, and updates of several worldwide stability guidelines are introduced. It highlights stability testing to support challenging storage environments in various climatic zones such as, South East Asia, India, or the Eastern Mediterranean regions. Specific challenges of scientifically designing stability programs for different types of pharmaceutical products are also discussed in this section.

The Technical Concepts for Stability Program Section explores several scientific aspects of a Stability Program. The application of the QbD concept to stability testing is examined. Safety and toxicology concerns of emerging impurities and stability of repackaged products or split tablets are discussed. This section also includes a review of physicochemical stability.

The Stability Data and Operational Practices Section covers concerns impacting stability data and regulatory submission. A discussion on setting specifications for active pharmaceutical ingredients (API) and various types of drug products is included. Approaches for maximizing the use of stability data to facilitate the approval of NDA/MAA and post-approval changes are presented. Finally, the strategies to build relationship with Contract Research Organizations (CROs) are evaluated, so that they become a critical extension of analytical research and development.

My hope is that this publication will serve those from industry, regulatory and academia well, whether they are practicing stability testing, developing stability program, monitoring quality systems, reviewing regulatory submission, or simply have a personal interest in this science.

Section II

Stability Studies in a Global Environment

Chapter 2

Regulatory Perspectives on Product Stability

Gary Buehler* and Kim Huynh-Ba

Abstract Recently, FDA has adopted the principles of Quality-by-Design and promotes the use of pharmaceutical development information in original applications. Stability studies occur in three phases in the life-cycle of a pharmaceutical product. They are: Stability studies during product development, stability studies that support a marketing application and stability studies that support post-approval changes. Quality-by-Design determines the stability studies conducted during product development. At the time of submission of an application, a package of stability data on the to-be-marketed product is combined with the knowledge gained during product development to establish a shelf-life for the drug product. After approval, routine stability data is collected to monitor product quality. When changes to the product formulation or its manufacturing occur, it may be recommended to the sponsor to obtain stability data to verify whether the change does not adversely affect the stability of the drug product.

1. Introduction

Stability is a critical issue in the pharmaceutical industry that affects the quality of drug products. Dr. Woodcock has indicated that "quality is free of contamination and reproducibly delivers the therapeutic benefit promised in the label to the consumer" (Woodcock 2004). Therefore, stability is defined as the consistent product quality and therapeutic benefit over the product's shelf life under various environmental conditions. Consumers expect that the products they take each day need to be stable and to deliver the correct amount of drug to produce the desired effect. Failures of stability lead to product recalls and may lead to loss of revenue for the sponsor. Table 2-1 lists the stability failures that led to recalls from 2004. It is the FDA's and manufacturers' responsibility to assure the product's stability.

For the twenty-first century, the FDA committed to the principles of Quality-by-Design (QbD) process, the goal of which is to ensure quality of drug products through their shelf life. Hence, the pharmaceutical product stability is

* The views expressed in this chapter are only of authors and do not necessarily reflect the policy of the U.S. Food and Drug Administration.

From: *Pharmaceutical Stability Testing to Support Global Markets*: Pharmasp,
Edited by: K. Huynh-Ba, DOI 10.1007/978-1-4419-0889-6_2,

Table 2-1. Examples of stability failures led to recalls from 2004 (tabulated in Gold Sheet, February 2005).

III	Stability failure (hydrocarbon)
II	Subpotent
II	Subpotent (stability)
II	Subpotent; product labeled as containing 1.0 mg/mL of morphine sulfate actually contains 0.2 mg/mL of morphine sulfate
III	Failed content uniformity testing
II	Subpotent; ethinyl estradiol component (18-month stability)
III	Superpotent
III	Subpotent; caffeine (9-month stability test)
II	Defective dropper: The ink graduated markings dissolve when stored in the product which may result in an inaccurate administration of prescribed dose
II	Superpotent; product exceeds in vitro enzyme levels (18-month stability)
III	Subpotent: stability
III	Superpotent: Super-potency of the active ingredient
III	Subpotent (Calcium Carbonate)
III	Superpotent; (18-month stability)
II	Superpotent; Selenium (12-month stability)
III	Subpotent: potential for tolnaftate solution to migrate into the swab tip during storage thereby producing inconsistent assay results
III	Subpotent; benzocaine (stability)

a function of the drug substance characteristics, formulation design, excipient selection, the container closure system and the manufacturing process.

Under QbD concepts, drug product stability studies verify the successful execution of product design, manufacturing process and the selection of the container closure system. They also provide evidence to support expiration dating periods and label recommendations for storage conditions. Therefore, evidence of acceptable stability is necessary to verify that the product continues to perform as expected to the expiration date. This evidence under the QbD mechanism occurs in three phases in the life cycle of a pharmaceutical product.

2. During Product Development

During product development, mechanisms of physical and chemical instability of the drug substance need to be defined. The sponsor must have a thorough understanding of the drug substance characteristics and identify external factors such as pH, light, temperature, and humidity. Interactions with excipients and packing materials must be well studied. The effect of manufacturing on stability is examined.

2.1. Mechanism of Instability

The mechanism of chemical degradation of the drug substance must be understood. Physical stability such as polymorphic form, dissolution profile and other quality attributes (e.g., resuspendability, aggregation, etc.) are considered during product development. Contributing factors such as pH, light, temperature and humidity are also very important to the drug product stability.

2.2. Interactions with Excipients

Excipient compatibility also plays a key role. The ICH Q8 guideline recommends this evaluation as part of the design of stable formulations. For solutions and suspensions, studies indicate which buffer to use and the optimum pH. For solids, compatibility studies are more extensive in order to select excipients to be used with the drug substances. In binary compatibility studies, excipients/active ingredient and excipients/excipients interactions are also very important.

During these formulation trials, design of experiments should be used to develop the formulation and to study stability of drug products.

2.3. Interactions with Packaging Materials

The container closure system can affect stability significantly. It directly links to the mechanisms of instability. The stability of the drug product is needed to be verified if the packaging configuration provides adequate protection to the drug product. For a liquid product, container orientation is also important as different degradation mechanism may result from a different storage orientation. In addition, leachability could create a subpotent product. Companies should consider a different orientation that may affect the drug product before it reaches the consumers.

2.4. The Role of Manufacturing Process

The manufacturing process must be selected based on the stability of the drug product. One must consider stability of the active ingredient within each process. For example, wet granulation would expose the API to solvents and high temperature. These exposures could be detrimental to an API that is highly sensitive to heat. Drug product stability must be evaluated to identify critical process parameters. Figure 2-1 illustrates how these factors contribute to product understanding. These factors could be designed based on Quality-by-Design concept, so that a firm could gain a thorough

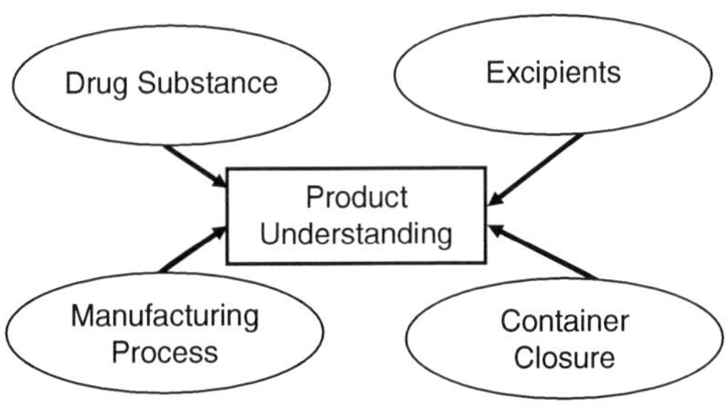

Fig. 2-1. Contributing factors to drug product.

understanding of the product and how it behaves through its shelf life. This understanding also helps the regulatory agency in evaluating the firm's application and expediting approval.

3. Support a Marketing Application

The QbD concept also helps NDA or ANDA submissions. It should be noted that the ICH Common Technical Document (CTD) format is recommended, and an electronic CTD is highly recommended. An electronic format will help the Agency review the application more efficiently. In fiscal year 2007, the Office of Generic Drugs (OGD) received approximately 900 abbreviated applications so that efficiency becomes critical.

4. Monitor Product Quality and Support Post-Approval Changes

Quality-by-design can be used to monitor product quality and also to support post-approval changes. After approval, quality of the drug product should be monitored routinely. When changes in manufacturing occur, additional data may be necessary to verify that the changes do not adversely affect the stability of the drug product. Therefore, the amount of stability data should be based on QbD concepts to assure the quality of the drug product is maintained through its shelf life.

5. Conclusion

On June 1, 2006, FDA withdrew the 1987 stability guidance and the 1998 draft stability guidance in anticipation of the Quality-by-Design approach. The pharmaceutical industry is interested in reducing the stability requirements, especially for post approval changes. As we know, product stability is a function of drug substance properties, formulation design, container-closure system, and the manufacturing process. Quality-by-design should be used to have a better understanding of these factors and to ensure product stability through expiry. The Agency strongly encourages the use of the QbD concept, but stability testing would continue to be part of a regulatory submission.

Reference

Woodcock J (2004) The concept of pharmaceutical quality. Am Pharm Rev 3:1–3

Author Biographies

Mr. Gary Buehler was appointed Director, Office of Generic Drugs, Center for Drug Evaluation and Research, in July of 2001. He had held the position of Deputy Director, Office of Generic Drugs, since May of 1999. Mr. Buehler has worked for FDA since 1986. Prior to joining the Office of Generic Drugs, he was a Senior Regulatory Project Manager in the Division of

Cardio-Renal Drug Products in the Office of New Drugs. Mr. Buehler retired from the United States Public Health Service (USPHS) in April of 2000. He served at a variety of duty stations in the USPHS Division of Hospitals and Clinics, as well as serving in the Indian Health Service in Nevada and Montana before coming to FDA. He graduated from Temple University School of Pharmacy.

Mrs. Kim Huynh-Ba is the Technical Director of Pharmalytik. She has over 24 years of experience in various analytical areas of pharmaceutical development and a primary focus in stability sciences, analytical development, outsourcing and technology transfer management. She has authored numerous technical publications and is invited frequently to present at national and international conferences. She is the editor of the "*Handbook of Stability Testing in Pharmaceutical Development: Regulations, Methodologies and Best Practices*", Springer.

Chapter 3

Current International Harmonization Efforts

Justina A. Molzon

Abstract Efforts towards international cooperation or harmonization can occur at varying levels – at the level of technical and scientific requirements, of dossier format and content, or of assessment and review practices. Three contrasting models for regional cooperation in drug regulation harmonization, the *corporate*, *populist*, and *cooperative*, are described and compared. Utilizing a specific case, drug stability testing requirements in Climatic Zone IV, the author explores each organizational model in action, reviews recent history of a key stability debate among harmonization initiatives from different climatic zones, discusses benefits and disadvantages seen in organizational models, and proposes the best practices approach to global harmonization efforts for the future.

1. Introduction

As the Associate Director for International Programs at the U. S. Food and Drug Administration (FDA) Center for Drug Evaluation and Research (CDER), I have the pleasure of serving as the CDER representative on both the International Conference on Harmonization (ICH) and Pan American Network for Drug Regulatory Harmonization (PANDRH) Steering Committees, and also as a member of the World Health Organization (WHO) Expert Committee-Specifications for Pharmaceutical Preparations. CDER's mission is to promote and to protect public health by assuring that safe and effective drugs are available to Americans. Congress passed the Food and Drug Administration (FDA) Modernization Act (FDAMA) in 1997, affirming CDER's public health protection role as it called for FDA to participate and collaborate, through appropriate processes, with representatives of other countries to reduce the burden of regulation, harmonize regulatory requirements, and achieve appropriate reciprocal arrangements. Based on my experience in working with various harmonizing initiatives, and the need to carry out Congressional legislation, this article provides my perspective on current international harmonization efforts.

From: *Pharmaceutical Stability Testing to Support Global Markets*: Pharmasp,
Edited by: K. Huynh-Ba, DOI 10.1007/978-1-4419-0889-6_3,

2. Harmonization in Drug Regulation: Three Models

The second Pan American Conference on Drug Regulatory Harmonization, held in Washington, D.C. in 1999, established "harmonization" as the delineation of common purpose among international drug regulatory authorities, identified within the framework of recognized standards and acknowledging differences in political, legislative, and health priorities among the region's countries. This definition speaks elegantly to international drug harmonization efforts to integrate national and international regulatory standards, facilitate universal acceptance among participating countries, and support efficient global drug development and local registration. Optimal harmonization can be promoted in terms of scientific and technical requirements, in the format and content of dossiers, and at the level of assessment and review practices – all pivoting on the cornerstones of safety, efficacy, and quality.

While technical and content requirements are the requisite elements of any harmonization program, international harmonization efforts, to date, have seen the emergence of more than one functional organization model. These models are premised on varying philosophies and are keyed to somewhat different approaches to issue prioritization, data gathering, problem-solving, and productivity. While no single harmonization model is innately superior to another, my observations and participation at "ground level" offer insights on the creation of enhanced organizational structures in the international harmonization efforts.

The International Conference on Harmonization of Technical Requirements for the Registration of Pharmaceuticals for Human Use (ICH) supports these objectives with guidelines on technical requirements for the registration of human drugs, as well as the use of the ICH Common Technical Document (CTD), which standardizes regulatory submissions within ICH countries and results in more efficient submission processes.

Three harmonization models are seen in action in the world, today. All bring unique strengths, but one, as an amalgam of the first two, may be seen as the "state of the art" in ensuring the best practices approach. The *populist* model features a flexible structure and open membership (as seen in the Pan American Network for Drug Regulatory Harmonization, or PANDRH). The *corporate* model, utilized by the ICH, is distinguished by restricted membership in the context of a closely managed formal structure. The *cooperative* model, exemplified by the World Health Organization (WHO), melds populist and corporate model assets in fostering collaboration, global partnerships, and "equal voice" initiatives, while still capitalizing on the benefit of structured practices, balanced oversight, and measurable outcomes.

While populist and corporate models each confers unique advantages, both tend to evolve toward the other over time, representing a structural convergence that organically evolves toward a cooperative model – which accommodates inclusionary practices and participatory balance with transparent processes and delineated workflow, all contributing to the overarching goal of harmonization: to make clearly visible contributions to regional and local health and well-being.

2.1. Populist Model: The Pan American Network for Drug Regulatory Harmonization (PANDRH)

2.1.1. The Establishment of PANDRH

The Pan American Health Organization (PAHO) promotes regulatory harmonization throughout the Americas, supporting improvements in the region's health by facilitating the availability of safe and effective pharmaceuticals. Regional harmonization also strengthens local drug regulatory authorities as well as the overall regulatory structure throughout Latin America.

PANDRH was established in 1999, at PAHO's second Pan American Conference on Drug Regulatory Harmonization. When a definition of "harmonization" was established at that meeting – finding common objectives within the framework of recognized standards while acknowledging differences in political, legislative, and health priorities, it was also understood that variabilities in harmonization would be expected due to varying capacities among PANDRH countries.

2.1.2. PANDRH Working Structure

Participants at PANDRH's biennial conference include regulators from the Americas (North, South and Central) along with the representatives from both innovator and generic pharmaceutical industry associations, regional economic groups, professional associations, consumer groups, and academia. Conference recommendations go to the PANDRH steering committee for feasibility evaluation prior to assignment to working groups consisting of regional representatives. These working groups utilize the recommendations in developing harmonized guidelines and training programs. A prioritization of seven key issues have been identified by PANDRH as *urgent*, *important*, and *recommended*. *Urgent* issues include Good Manufacturing Practices (GMP), Bioavailability/Bioequivalence (BA/BE), Good Clinical Practices (GCP), and Counterfeit Pharmaceuticals. *Important* issues are classification (generic, OTC, etc.) and drug regulatory authorities assessment and function. The *recommended* issue is pharmacopoeia standards evaluation.

PANDRH's organizational model, characterized by open membership, flexible structure, and inclusivity, functions to promote guideline implementation, but also makes consensus across a wide collective of stakeholders more difficult to reach. This aspect of the populist model, in turn, results in nebulous objectives and a focus on capacity-building and training over well-defined deliverables and managed outcomes.

2.2. International Conference on Harmonization

2.2.1. ICH Background

The ICH was established in 1990 as a partnership between the drug regulatory authorities and pharmaceutical industries in Europe, the United States, and Japan (where a majority of pharmaceutical research and development occurs). From its inception the ICH sought to minimize or eliminate duplication in testing and analysis during research and development of new drugs by identifying, assessing, and recommending harmonized principles in the application of technical guidelines and requirements for product registration. Such harmonization has been clearly shown to reduce unnecessary delay in the development and availability of new medicines, and maintains quality and efficacy while also maintaining regulatory obligations to protect public health and safety.

To date, the ICH has developed and disseminated over 50 different technical guidelines.

2.2.2 ICH Keys of Success

The success of the ICH is based on defined processes, effective administration, a limited number of players with a common focus driven by comparable regulatory, technical, and financial capacities, as well as a commitment by participating members to implement ICH-approved guidelines. With its management-oriented organizational structure, systematic operational principles, and prescriptive guidelines, ICH represents a *corporate* model. As a member-funded organization with a dedicated secretariat, ICH is particularly well-positioned to maintain sustainable efforts and measurable outcomes.

2.3. ICH Global Cooperation Group

2.3.1. Establishment of the Global Cooperation Group

Despite ICH's initial focus on behalf of Europe, the United States, and Japan, interest quickly grew in ICH activities from non-ICH countries and regions, all of whom might benefit from improved access to technical data and registration guidelines. In this context, it also became apparent that ICH guidelines must be applied to all manufacturers (for example, generics and OTC formulations), or risk a two-tiered approach to quality and efficacy that, by definition, served to undermine the concept of harmonization.

Responding to these issues, ICH formed the Global Cooperation Group (GCG) in 1999, operating as a subcommittee of the ICH Steering Committee, with membership comprised of one representative from each of the six parties on the ICH Steering Committee, the ICH Secretariat, and two Observers (WHO, Canada).

Initially, the GCG functioned only as a clearinghouse, offering information on ICH, ICH activities, and ICH guidelines to any country or company that requested information.

While the GCG's mission from the outset was to promote mutual understanding of ICH guidelines by regional initiatives and facilitate capacity of drug regulatory authorities to implement those guidelines, it became clear that a more active engagement was needed to respond not simply to burgeoning interest in ICH processes by regional harmonization initiatives and industry, but to accelerate interest in *participation* in those processes: granting input, contributing to technical concept discussions and guideline development, and the opportunity to represent and communicate issues with critical bearing on regional harmonization, including environmental, economic, and geopolitical factors that are unique to regions outside of ICH's initial EU-US-Japan partnership.

In its evolution from information clearinghouse to a partnership, the GCG has evolved toward a cooperative model, a blend of the populist and corporate models that adopts a "best of both worlds" stance, creating inclusive yet structured processes that ensure transparency, balanced communication, and mutual engagement in promoting the "best practices" approach to the implementation of ICH guidelines among all stakeholders, in the context of their specific national priorities.

3.3 Global Stability Testing Requirements: A Case Study in Models of Harmonization

Efforts toward harmonized drug stability requirements offers a striking illustration of the three models of harmonization at work in addressing the critical "real world" problem of drug stability in multiple different climatic zones.

Table 3-1. Grimm's climatic zones and recommended storage conditions.

Climatic zone	Definition	Storage conditions
I	Temperate	21°C/45% RH
II	Subtropical; Mediterranean	25°C/60% RH
III	Hot, dry	30°C/35% RH
IV	Hot, humid	30°C/70% RH

Drug stability involves numerous aspects of the research and development process, including formulation, analytical testing, toxicology, and quality – all of which have far-reaching impacts on regulatory concerns, from drug synthesis to formulation compatibilities to approval criteria and supply chain planning. Stability testing applies to all batches of a product across multiple criteria, with resulting data expected to confirm that a product meets acceptance parameters throughout a designated shelf life, and is appropriate for regulatory approval and registration. As such, drug stability is an integral aspect of the overall assurance of medicinal integrity.

The effect of climatic conditions on drug stability was investigated by Paul Schumacher, who first proposed an approach to drug stability through the designation of climatic zones in 1972. Wolfgang Grimm expanded on Schumacher's ideas in 1986, dividing the planet into four climatic zones based on the mean annual temperature and humidity. This conceptual format provided a metric for drug storage conditions based on temperature and relative humidity. In other words, a medicine, which can be of any form (tablet, capsule, liquid) must maintain stability (pharmacotherapeutic activity) under specified climatic conditions for a specified period of time (see "Storage Conditions" in the table below). Table 3-1 below presents Grimm's original schematic for the assessment of climatic zones and their implications for drug stability.

Grimm's framework informed the development of the ICH's "Q1F" guideline (relevant to climatic zones III and IV.) Q1F, adopted by the ICH in February 2003, defined storage conditions for stability testing in countries not located in the ICH regions and not covered by the ICH's "master guideline" for stability testing of new drug substances and products (Q1A (R2)). The ICH's initial goal was straightforward: harmonized global stability testing requirements through the reduction in the number of different storage conditions. 30°C/65% RH was defined as the long-term storage condition for Climatic Zones III/IV. Q1F reviewed storage conditions for countries and cities in these climatic zones, and offered guidelines regarding container closure systems and shelf life/product expiration labeling.

From the outset, the "hot and humid" regions of Zone IV was a key issue. Debate quickly ensued regarding proper temperature and humidity parameters to protect drug product quality in Climatic Zone IV, notably from the Association of Southeast Asian Nations (ASEAN) Pharmaceutical Product Working Group, who wanted to include a larger safety margin for medicinal products intended for marketing in their region.

The ICH subsequently reported that "several countries and regions ... revised their own stability testing guidelines, defining up to 30°C/75% RH as the long-term storage conditions for hot and humid regions." What the ICH faced was a tension between its corporate model of harmonization and the more populist model as embodied by ASEAN. More to the point was the fact that Zone IV ASEAN Member States made it clear that, in their view, being neither invited nor involved in ICH stability discussions and guideline development implied that they were under no obligation to follow those guidelines – but could, instead, develop and adopt their own guidelines specific to their own climatic realities.

Zone IV Regional Harmonization Initiatives (RHIs), with their own unique set of public health responsibilities, found themselves involuntarily defined by the ICH's corporate model of harmonization, offered guidelines that were reviewed and approved with, in their view, inadequate input or consensus-building.

The WHO 40th Expert Committee on Specifications for Pharmaceutical Preparations (October 2005) would initially address this concern with recommendations that Zone IV guidelines be amended as Zones IV-A and IV-B (30°C/65% RH and 30°C/75% RH, respectively), and that each individual Member State within the former Zone IV should indicate which of these two conditions would be applicable in its territory. A key objective of this recommendation was to have countries formally declare their climatic status so that industry would know the stability requirements much earlier in the drug development process, in view of the many years involved, as well as the formidable extent of testing and analysis required in drug stability studies.

These recommendations were expanded on at the WHO-ICDRA (International Conference of Drug Regulatory Authorities) Conference in April 2006, where I chaired a workshop on global challenges in harmonization of stability testing issues. That workshop generated further recommendations in the service of an ongoing, balanced, and inclusive discussion:

- Member States should identify their stability testing conditions in order to facilitate import to / export from their country.
- Member States should make information available to WHO regarding stability conditions to be used within their markets.
- WHO should make available Member States' information in order to facilitate access by manufacturers and any interested party on an international basis.
- WHO should observe the situation and any future developments and continue its efforts to identify harmonized conditions.
- Any international mechanism or organization which develops guidance relevant for countries outside their own regions should ensure that those countries are made aware of developments and are directly approached to participate in the consultation process. For the ICH, the GCG should be emphasized as a pathway to collaboration with regional harmonization initiatives (RHIs).

In June 2006 the ICH reported that "due to this divergence in global stability testing requirements, the ICH Steering Committee has decided to withdraw ICH Q1F," electing to leave definition of storage conditions in Climatic Zones III and IV to the respective regions and WHO. At the time of guideline withdrawal, ICH noted in the official communications that "in assessing the impact of the withdrawal of ICH Q1F on intermediate testing conditions defined in ICH Q1A (R2), the decision was reached to retain 30°C/65% RH. However, regulatory authorities in the ICH regions have agreed that the use of more stringent humidity conditions such as 30°C/75% RH will be acceptable provided the applicant decide to use them."

With that, ICH recognized the separation of Climatic Zone IV into the two subdivisions endorsed by WHO eight months earlier, and by so doing it had been indirectly acknowledged that a process of discussion and collaboration (i.e., a populist model) had merit, if not necessity, in bringing consensus to harmonization as it also more effectively addressed the specific needs of RHIs not originally included in the ICH regional mandate.

4. Conclusion

This case report of stability testing in the contexts of different harmonization models both illustrates and underscores my larger thesis that a fusion of corporate and populist models, creating a cooperative model, presents the "best practices" avenue forward in global harmonization efforts. We have seen how a strictly corporate model, while thorough and based on strong science, tended to exclude key stakeholders from the very processes and decisions most germane to their work in regulating drug products and assuring a safe supply of quality medicines for their populations. This exclusion led to contention and dissatisfaction – as opposed to "harmonization" – while also adding multiple additional administrative layers to the process, eliciting industry concern, and frustrations regarding efficient information flow among all stakeholders.

By the same token, the populist model can often find itself mired in local interests and organizational barriers to reach consensus. In the service of accommodating all voices and concerns, the path forward to guideline finalization and approval in a populist organizational model can be lengthy, or, at worse, delayed for extended periods of time.

A cooperative model tends to ameliorate the disadvantages of both corporate and populist models, while amplifying the advantages of both. In the case of drug stability consensus, guidelines, and implementation, as well as enhanced and accessible information for industry, it is evident that a cooperative model of harmonization represents a current best practice modality.

The ICH GCG is a current forum through which open discussions, partnerships, and collaborative decision-making can flourish. With the GCG's policies of open information dissemination as well as its underlying principle to not impose views or opinions in the absence of discussion and negotiation, it represents a standing alternative to dissension and, within the ICH's larger construct, a cooperative model in action.

Key conclusions in marking a way forward in global harmonization initiatives include:

- Regulators from Zone IV countries should be included in all discussions and decision-making processes.
- Utilization of the ICH GCG is recommended to discuss differences, engage multiple opinions, elicit expert guidance, reach consensus – and capitalize on the benefits of a cooperative model of harmonization.
- Collaboration helps to resolve differences and to enhance global harmonization efforts, leading in turn to more efficient regulatory systems and increased availability of safe, effective, quality pharmaceutical products on a global level.

Author Biography

Capt. Justina A. Molzon, a pharmacist and attorney, is a commissioned officer in the Public Health Service. She is currently the Associate Director for International Programs, Center for Drug Evaluation and Research (CDER), U.S. Food and Drug Administration and a member of CDER's senior management team. One of her primary responsibilities is coordination of CDER's efforts related to the International Conference on Harmonization of Technical Requirements for Registration of Pharmaceuticals for Human Use (ICH). In this capacity, she serves as CDER's representative to the ICH Steering Committee. She received her B.S. (with honors and distinction) and M.S. in Pharmacy, with a concentration in pharmaceutics and pharmacognosy, from the University of Rhode Island and her law degree from the Chicago-Kent College of Law/Illinois Institute of Technology. She is a Fellow in the American Society for Pharmacy Law, serving on its Board of Directors from 1989 to 1997. Justina served as a pharmacist in the Public Health Service on the Navajo Indian Reservation and also served as the Regional Pharmacist Consultant and Inspector for the Health Care Financing Administration's Survey and Certification Review Branch, in Chicago. Upon leaving the Public Health Service, she became a clinical pharmacist in the critical care area of Northwestern University's teaching hospital. While in Chicago, she also maintained a private law practice and worked with a pro bono legal program for persons with AIDS. In 1990, Justina rejoined the Public Health Service to work in FDA's Office of Generic Drugs. She is very grateful that she has the opportunity to work with international drug regulatory authorities in assuring the safety, efficacy and quality of pharmaceuticals worldwide.

Chapter 4

Update on the WHO Stability Guideline

Sabine Kopp

Abstract The work on stability was initiated by the World Health Organization (WHO) in 1988. Following the WHO process for consultation a general text on stability and the *WHO Guidelines on stability testing for well established drug substances in conventional dosage forms* were adopted in 1994 and 1996, respectively. In 2000, discussions were initiated between the International Conference on Harmonisation (ICH) Expert Working Group Q1 (stability) and WHO in order to harmonize the number of stability tests and conditions undertaken worldwide. Discussions on the stability testing conditions for climatic zone IV (hot and humid) have been taken up internationally. Based on the various stability-related guidances published by national authorities, as well as regional harmonization groups, discussions are still ongoing and have triggered various changes to the WHO guidelines on stability testing. Based on the recommendations by the International Conference of Drug Regulatory Authorities and the abovementioned WHO Expert Committee, a revision of the WHO guidelines is currently underway.

1. Introduction

The main functions of the World Health Organization (WHO) were established by the Organization's constitutional mandate. These are to act as the directing and coordinating authority on international health work, to encourage technical cooperation for health with WHO Member States and to "*develop, establish and promote international standards with respect to biological, pharmaceutical and similar products*" (Article 2(u) of WHO's Constitution). WHO's technical advisory body for normative work related to medicines is the WHO Expert Committee on Specifications for Pharmaceutical Preparations. The recommendations of this Committee are communicated to the Executive Board and the World Health Assembly, and thus to all WHO Member States.

The World Health Organization has 193 Member States, which has its annual meeting in Geneva. In 1948, the WHO Expert Committee on Specifications for Pharmaceutical Preparations was formed to discuss quality issues of medicines.

From: *Pharmaceutical Stability Testing to Support Global Markets*: Pharmasp,
Edited by: K. Huynh-Ba, DOI 10.1007/978-1-4419-0889-6_4,

The revision of the WHO guidelines on stability testing published in 1996 is currently being revised. The first draft of the new WHO stability guidelines was circulated for comments in April 2007 . These comments were discussed during a consultation held in June 2007. The second draft was made available in October 2007 based on the WHO Eastern Mediterranean Region stability guidance.

2. Basis for WHO International Guidelines

2.1. WHO Expert Committee

The WHO Expert Committee is an official advisory body, governed through the Organization's rules and procedures. Expert Committee members are drawn from the WHO Expert Advisory Panels. In addition to these officially designated WHO experts, there are also technical advisers and observers from international organizations, institutions and nongovernmental organizations (NGOs). Participation in the Expert Committee meetings depends on the topics of discussion in order to assure applicability of guidelines. Based on these discussions, the members of the WHO Expert Committee issue a report to recommend adoption of any guidelines that were discussed and circulated prior to the meeting on a global level. These recommendations are presented to the Director-General of WHO and to a Governing Body of WHO, the Executive Board, before their publication and subsequent implementation by WHO Member States. Although the process is complex, it is important that the correct procedure is followed to ensure that the new WHO guidelines are scientifically sound and internationally acceptable.

2.2. WHO Processes

Before starting the development of any new WHO guideline, the topics for the future work are identified, either by the WHO Governing Bodies, WHO Members States or related bodies. GMPs, for example, are developed by the WHO Expert Committee on Specifications for Pharmaceutical Preparations upon request by the World Health Assembly. The consultation process is global (see Fig. 4-1) and involves numerous partners as described below. Once a working draft is deemed globally acceptable, then the guideline can be adopted and published. This process can take up to 7–8 years, as in the case of stability.

2.3. WHO Partners

WHO's partners include national and regional authorities, international organizations, professional associations, institutions, NGOs, etc. WHO also works with pharmaceutical industry associations such as the International Federation of Pharmaceutical Manufacturers and Associations (IFPMA), International Generic Pharmaceutical Alliance (IGPA), and the World Self-Medication Industry (WSMI). Partners also comprise specialists from all quality assurance related areas, including regulatory, university, and industry.

Then there are the WHO Collaborating Centers with which the Organization works, e.g., for validation of the written standards in the laboratories. WHO also works closely with regional and interregional harmonization groups like ICH.

Step 1. Preliminary consultation and drafting

Step 2. Draft guidelines

Step 3. Circulation for comments

Step 4. Revision process

.... (back to Steps 2 and 3 as often as needed)

→ WHO Expert Committee (EC) meeting

→ if guideline adopted, published in EC report as an Annex

-> *if not, back to Steps 2-4*

-> WHO Governing Bodies

-> Recommendation to Member States for implementation

Fig. 4-1. System to discuss WHO guidelines.

3. Update of WHO Guidelines

The WHO Expert Committee on Specifications for Pharmaceutical Preparations covers the area of quality assurance of medicines from production through to distribution, i.e., from manufacture to delivery to the patient. The Committee started its work in the area of stability in 1988. It took quite a long time, namely 8 years, until this work was finalized, mainly due to a large process of consultation and observing experts' advice as to what was happening in the ICH environment. This was considered necessary as the experts did not want to develop a different set of stability testing conditions.

The WHO stability guidelines were, at that time, focusing on pharmaceutical products and those that were well-established, i.e., generic products, in conventional dosage forms, as this was considered to be a priority.

The world was divided into four zones:

- *Zone I:* temperate
- *Zone II*: subtropical, with possible high humidity
- *Zone III*: hot/dry
- *Zone IV*: hot/humid

The most critical zones are the hot-dry and hot-humid conditions. The storage conditions are explained in several references and calculated data (Schumacher 1972; Grimm 1986; Zahn et al. 2006; Huynh-Ba 2008).

In 1996, the testing condition chosen for Zone IV was 30°C/70% RH. In 2000, there was a request from ICH to modify the standard condition for Zone IV. This proposal, to change from 30°C/70% RH to 60% RH, was submitted to the WHO consultation process, but was not found to be acceptable. In 2001, a new proposal was received to modify both the ICH and the WHO guidelines for Zone IV to 65% RH instead of 70% RH. Again, this proposal was widely circulated for comments and found acceptable by most experts. Thus the WHO testing conditions for Zone IV were modified to 65% RH. Some Member States raised the concern that this condition would not be applicable in their country. Therefore, the experts included a provision for transportation and storage conditions when outside these criteria. In 2003, the ICH Stability Data Package for Registration Applications in Climatic Zones III and IV (Q1F) was signed off by all ICH partners and the conditions were found to be in line with those discussed by the aforementioned WHO Expert Committee. In 2004, a number of meetings were held in the Association of South-East Asian Nations (ASEAN) region to discuss its stability testing conditions. In January 2004 new conditions, based on real methodological data, were proposed by ASEAN. These new conditions for Zone IV were 75% RH. Many discussions were subsequently triggered at an international level and because of that, these new conditions have been found accepted in the other parts of the world, e.g., in Brazil. In October 2004, the WHO Expert Committee meeting recommended further discussion at an international level because the so-called Zone IV is now consisted of two conditions. In December 2004, a meeting was organized by WHO in Geneva. The outcome was of the proposal three options and a plea to all the WHO Member States and all interested parties to express which of the three conditions they would find acceptable. In October 2005, the Expert Committee reviewed all feedback, and discussed and recommended to go with two different Zones IV, i.e., Zone IVA, IVB, in order to avoid creating a third set of conditions at the WHO level. Each WHO Member State was asked to indicate which conditions would be applicable in its territory.

Further developments in 2006 saw ICH withdrawing its Q1F. The guideline was withdrawn due to the divergence in global stability testing requirements and the definition of the storage conditions in climatic zones III and IV. It was left to the respective regions and to WHO to define the respective stability testing conditions for those regions. It was also decided that the intermediate conditions would be retained in order to facilitate the harmonization process and to avoid another set of conditions.

The International Conference on Drug Regulatory Authorities (ICDRA) key recommendations on stability testing made during its 12th meeting in 2006 is included:

- WHO Member States should identify their stability testing conditions to facilitate import to and export from their countries.
- WHO Member States should make information available to WHO regarding stability conditions to be applied within their markets.
- WHO should make available country information to facilitate its accessibility to manufacturers and any interested party on an international basis.

4. Key Recommendations on Stability Testing

WHO corresponds with its 193 Member States and major regional harmonization groups requesting information on long-term stability conditions. The new WHO stability guidelines were prepared in accordance with the usual procedure, through wide circulation for comments, collation of comments and discussion during WHO consultations.

This new proposal, i.e., the current revision, will include both active pharmaceutical ingredients, as well as finished pharmaceutical products; thus the scope of the guidelines is expanding. The underlying concept is to accommodate the intended market, which means three main long-term conditions corresponding to Zone II, Zone IVA and Zone IVB. The guidelines will also include different scenarios, a general case and refrigerator and freezer storage, which were not included in the previous guidelines. There are also recommendations on labeling derived from stability studies. They will further include the case of semi-permeable and impermeable containers and a determination of water loss, stress testing in more detail than previously and the stability commitment. This second draft gives preference to the real conditions required by national authorities when compared to the first draft, where the climatic conditions were given as per the references in the literature due to the numerous remarks and comments received from some national drug regulatory authorities. The experts, therefore, decided that it was better to focus on the conditions that are actually requested rather than providing a list of countries with their climatic zones, especially as some countries contain several zones within their area.

5. Conclusion

To summarize, the WHO stability guidelines are under revision. They are currently foreseen to cover marketed and new pharmaceuticals, active pharmaceutical ingredients and finished pharmaceutical products, and will comprise an annex with long-term conditions required for marketing authorization in WHO Member States.

References

Grimm W (1986) Storage conditions for stability testing (Part 2). Drugs Made Ger 29:39–47

Huynh-Ba K (ed) (2008) Handbook of stability testing in pharmaceutical development: regulations, methodologies and best practices. Springer, New York, pp 43–91

Schumacher P (1972) Über eine für die Haltbarkeit von Arzneimitteln maßgebliche Klimaeinteilung. Pharm Ind 34:481–483

Zahn M et al (2006) A risk-based approach to establish stability testing conditions for tropical countries. J Pharm Sci 95:946–965

Author Biography

Dr. Sabine Kopp is Secretary of the WHO Expert Committee on Specifications for Pharmaceutical Preparations and is responsible for the normative work related to Quality Assurance of Medicines within WHO. She received her Ph.D. in Pharmaceutical Technology from the University of Tübingen (Germany) in 1986. Dr. Kopp has over 20 years' experience in international pharmaceutical quality assurance standard setting. She has worked in various positions within the World Health Organization, including Secretary for the International Nonproprietary Names (INN) Programme, Acting Team Coordinator for Quality Assurance and Safety of Medicines, in the Department of Essential Drugs and other Medicines. Dr. Kopp coordinated work and related WHO publications for the INN Programme for pharmaceutical substances, *The International Pharmacopoeia*, including related quality control matters, and guidance on quality assurance, such as good manufacturing practices (GMP), guidelines on stability testing and interchangeability of generic medicines and inspection. She has taken part in WHO projects on the implementation of GMP and on Counterfeit Drugs. Dr. Kopp presents at and represents WHO in, international fora, including the Pharmacopoeial Discussion Group (PDG), International Conference on Harmonisation (ICH), Panamerican Conference on Drug Regulatory Harmonization, the World Bank, World Customs Organization (WCO) and the World Intellectual Property Organization (WIPO).

Chapter 5

Development of a Regional Guideline for the Eastern Mediterranean Region

Abdel Aziz Saleh and Kim Huynh-Ba

Abstract The World Health Organization/Eastern Mediterranean Regional (WHO/EMR) Office has organized a consultation on "Regional Guidelines on Stability Studies of Medicines and Biologicals" on February 25–28, 2006, in Jeddah, Saudi Arabia. The meeting was attended by participants from 12 countries from the region. The objectives of this consultation were to: (a) review national requirements for stability studies of medicines and biologicals; (b) update participants with recent advances in the area of stability studies of medicines and biologicals; and (c) develop regional guidelines on stability studies of medicines and biologicals. In September 2006, the WHO/EMRO Regional Committee adopted the regional guidelines on stability testing of active substances and pharmaceutical products after the inclusion of comments made by the member states as appropriate and requested member states to revise the requirements for drug registration to be consistent with these guidelines. The proposed guidelines constitute the basis for the revision of the Global WHO Stability Guidelines. Several other recommendations were also passed to emphasize the need to strengthen national expertise in this area and to carry out post-marketing studies to ensure drug stability throughout the distribution channel and the recommended shelf-life.

1. Historical Background

The World Health Organization (WHO) Eastern Mediterranean Region (EMR) is composed of 22 countries, most of which are Arab countries together with Afghanistan, Iran and Pakistan.

In 1993, the Regional Office organized an Inter-country Workshop on the Validation of Expiry Dates of Drugs in Amman, Jordan to form the basis for guidance on stability testing of pharmaceuticals in the region consisting of 32 countries. The workshop recommended that the medicine regulatory authorities, in close collaboration with pharmaceutical industry, professional associations and academic institutions, develop clear guidelines on requirements for drug stability and expiry date determination. Thereafter, extensive

From: *Pharmaceutical Stability Testing to Support Global Markets*: Pharmasp,
Edited by: K. Huynh-Ba, DOI 10.1007/978-1-4419-0889-6_5,

discussion was done and the concept of stability guideline was introduced. Consequently, the Arab Union of Pharmaceutical Manufacturers in collaboration with WHO published the Arab Guidelines on Stability Testing of Pharmaceutical Products. In 1995, the *Arab Guidelines on Stability Testing of Pharmaceutical Products* were approved by the Ministers of Health of the Arab for stability testing. This guideline was well written but the implementation was not well ministered. After several years, information collected from various countries of the region showed that most of them are not following the regional guidelines. Stability requirements for different Arab countries were not consistent as well as not clear. Most of the Arab countries neither had a requirement for the stability to exercise, nor clear labeling of shelf life and storage conditions. In 2003, the Executive Board of the health ministers' council for the Gulf States met and issued the GCC[1] Guidelines on Stability Testing of Pharmaceutical Products. Similar to the previous guidelines, this document did not get a good entry to the countries of that region. In February 2006, the Consultation on Regional Guidelines on Stability Studies of Medicines and Biologicals was held in Jeddah, Saudi Arabia (Saleh 2007).

2. General Aspects of Medicine Stability

The following are the key concepts for Medicine Stability. National medicine regulatory authorities should clearly indicate the national stability requirements and should have the national expertise to validate the stability data provided, as well as monitor the quality of medicine throughout the shelf-life. In this respect, harmonization of stability requirements at regional and global levels is of great importance.

2.1. Educational Aspect

Education is the most important issue. Pharmacy students should be taught, and pharmacy curriculae should include the scientific principles and practical aspects of stability of medicinal products, their recommended storage conditions and expiry dates. Such courses should include the practical aspects related to real storage conditions. This clearly indicates the importance of teaching the concept of mean kinetic temperature and the actual temperature and humidity conditions during storage of medicinal products.

2.2. Operational Research

To ensure stability throughout the shelf-life, it is necessary to carry out field research on quality of medicines stored under practical conditions in each country. The results of such studies would provide important information to regulatory bodies on the national requirements for stability studies and storage conditions. A search of the regional research data base, the EMR Index Medicus, shows very few research studies have been conducted in this area.

[1]The Cooperation Council for the Arab States of the Gulf.

Table 5-1. Guidelines presented by members within the region.

- *Sultante of Oman:* essentially follows the GCC requirements
- *Jordan:* Although the majority of the population live in a Mediterranean climate, i.e. long-term testing at 25°C/60% RH would be adequate, the authorities require data generated at 30°C/65% RH to cover the hot and dry desert
- *Drugs Control in Morocco* by National Laboratory of Drug Control (N.L.D.C) Rabat. Long-term testing required at 25°C/60% RH (CZ II) or at 30°/65% RH
- *Pakistan:* Long-term testing at "room temperature", accelerated condition 37–40°C/75% RH for 6 months
- *The Saudi Food & Drug Authority (SFDA) Guidelines for Stability Testing of New Drug Substances and Products* adopted ICH Guideline Q1A for CZ IVA
- *Registration of Pharmaceutical Products in Sudan* by the drug quality control laboratories require stability data at 30°C/65% RH for at least 24 months and at 40°C/75% RH for 6 months
- *Egypt:* Testing conditions adopted from ICH Guideline Q1A
- *Kingdom of Bahrain:* ICH Guideline Q1A essentially adopted
- *United Arab Emirates:* belong to CZ IVA, follow ICH Guideline Q1A

2.3. In-Use Study

In addition to operational research studies to validate the indicated expiry date under practical storage conditions, research should also include stability of medicines following reconstitution and storage before use. Expiry dates following reconstitution should also be validated.

In addition, patients should be clearly advised on how to store the medicines properly, particularly reconstituted medicines. Patients should be advised to observe expiry dates of medicines, both those freshly procured and those stored in the house, and they should be advised on what to do with the expired medicines. Information on storage conditions, and expiry dates should be clearly labeled on medicines in national languages.

3. Countries Situation

At the consultation meeting in Jeddah in February 2006, regulatory requirements on the stability of medicine in different parts of the region were presented. It became obvious that there was a large difference among the region members. Table 5-1 lists the guidelines presented by several countries.

4. Draft Regional Guideline

A draft regional guideline on stability testing of active substances and pharmaceutical products was developed by the Consultation on Regional Guidelines on Stability Studies of Medicines, in Jeddah, Saudi Arabia, on February 25–28, 2006. It was based in part on existing guidelines developed by ICH, the European Medicines Agency (EMEA), and the Cooperation Council for the Arab States of the Gulf (GCC). The guidelines are intended to define the stability data packages for active substances and pharmaceutical products that are sufficient for the application for registration within countries of the Region. The guidelines address the information required for the applications for registration of New Chemical Entities, as well as existing active substances and their related pharmaceutical products for human use. In order to identify adequate testing conditions, the climates in countries of the Eastern Mediterranean Region were analyzed using climatic data extracted from the European Centre for Medium-Range Weather Forecasts. For recommended long-term testing condition, the hottest and most humid part of each country was selected to establish the adequate stability testing condition. In some countries, aqueous-based solutions in semi-permeable packaging, and dosage forms sensitive to low humidity, e.g., hard-gelatine capsules, may require testing at low humidity according to the procedure described in the guideline (Zahn 2006).

The purpose of the stability study is to establish, based on testing a minimum of two or three batches of the pharmaceutical product, a shelf life and label storage instructions applicable to all future batches of the pharmaceutical product manufactured and packaged under similar circumstances. The degree of variability of individual batches affects the confidence that a future production batch will remain within the specifications throughout its shelf life.

For products that are expected to be stable at standard conditions in Climatic Zone II, the stability study should be done at 25±2°C and 60±5% RH with a minimum time period covered by data at submission being 12 months. And for intermediate study, the storage condition should be 30±2°C and 65±5% RH with a minimum time period covered by data at submission being 6 months. For accelerated study, the storage condition should be 40±2°C and 75±5% RH with a minimum time period covered by data at submission being 6 months.

For products to be stored in refrigerator, long term study should be done at 5±3°C with a minimum time period covered by data at submission being 12 months, and for accelerated study the storage condition should be 25±2°C and 60±5% RH or 30 ± 2°C and 65 ± 5% RH, determined by the climatic zone in which the active substance is intended to be stored, with a minimum time period covered by data at submission being 6 months.

For products to be stored in freezer, long term study should be −20±5°C with a minimum time period covered by data at submission being 12 months. In the case that available long-term stability data on primary batches do not cover the proposed shelf life granted at the time of approval, a commitment should be made to continue the stability studies after approval in order to firmly establish the shelf life.

A systematic approach had to be adopted in the presentation and evaluation of the stability information, which should include, as appropriate, results from the physical, chemical, biological and microbiological tests, including particular attributes of the dosage form. For statement and labeling, a storage statement should be established for the labelling based on the stability evaluation of the pharmaceutical product, where applicable, specific instructions should be provided, particularly for pharmaceutical products that cannot tolerate freezing. Terms such as "ambient conditions" or "room temperature" must be avoided. The purpose of in-use stability testing is to establish, where applicable, the period of time during which a multi-dose product can be used while retaining acceptable quality, once the container is opened and the first dose is removed.

5. Recommendations

Several recommendations are made based on this document:

5.1. Technical Recommendations

It is recommended that Regional Committee Resolution adopts the regional guidelines on stability testing of active substances and pharmaceutical products following the inclusion of the comments made by Member States as appropriate. It is also requested Member States to revise the requirements of drug registration to be consistent with these guidelines. It requested WHO to revise the global WHO stability guidelines taking into consideration the regional guidelines and requested the Regional Director to continue efforts in the area of harmonization in other technical registration requirements, particularly those related to biogenerics and other biotechnology products.

5.2. Other Recommendations

It is recommended that the pharmacy schools have to revise the under-graduate and postgraduate courses to include basic content of the Regional Guidelines, as well as stability studies for biotechnology and biological products.

In addition, pharmacy schools are recommended to develop and implement regular training courses on stability studies and storage conditions. Experts at national pharmaceutical manufacturers are encouraged to arrange regular training courses on stability testing and storage conditions.

5.3. Storage Conditions

To assure the quality of the medicine during storage, it is recommended that manufacturers clearly indicate the appropriate storage conditions in national language, and implement Good Storage Practice (GSP) and Good Distribution Practice (GDP). Table 5-2 lists the climatic zones assigned to countries of the Region and their recommended long-term testing conditions. It would also be beneficial to develop and implement national public education courses on good storage practices. Pharmacy schools and research institutes should carry out operational stability research and studies to validate marketed pharmaceutical product quality throughout the shelf-life.

6. Conclusion

The above recommendations are necessary to assure the quality of the medicine remaining throughout the shelf life. Appropriate storage conditions will help to ensure Good Storage Practice. Public education program and cooperation between schools and health authority and health institutes will enhance pharmaceutical product quality. Current efforts are ongoing to focus on the implementation and development of sectors, including cooperation from the industry, and multinational and local manufacturers.

Table 5-2. Climatic zones assigned to countries of the region and their recommended long-term testing conditions.

Country	CZ II	CZ III	CZ IVa	Recommended long-term testing condition
Afghanistan	+	+		30°C/65% RH
Bahrain			+	30°C/65% RH
Djibouti			+	30°C/65% RH
Egypt	+	+		30°C/65% RH
Iran, Islamic Republic of	+	+	+	30°C/65% RH
Iraq		+		30°C/35% RH
Jordan	+	(+)		30°C/65% RH
Kuwait			+	30°C/65% RH
Lebanon	+	(+)		25°C/60% RH
Libyan Arab Jamahiriya	+	(+)		25°C/60% RH
Morocco	+			25°C/60% RH
Oman		(+)	+	30°C/65% RH
Pakistan	+	+	+	30°C/65% RH
Palestine	+			25°C/60% RH
Qatar			+	30°C/65% RH
Saudi Arabia		+	+	30°C/65% RH
Somalia			+	30°C/65% RH
Sudan		+	+	30°C/65% RH
Syrian Arab Republic	+	(+)		25°C/60% RH
Tunisia	+	(+)		25°C/60% RH
United Arab Emirates		+	+	30°C/65% RH
Yemen	+		+	30°C/65% RH

+ Climatic zone assigned
(+) Deserted part of the country

References

Saleh A (2007) Regional guidelines on the stability testing of active substance and pharmaceutical product for the Eastern Mediterranean region, presented at the AAPS, Workshop in Bethesda, MD, 10–12 Sept 2007

Zahn M (2006) A risk-based approach to establishing stability testing conditions for Northern Africa, the Arabian Peninsula, the Middle East and Pakistan, presented to the WHO Consultation on Stability Testing in Jeddah, 25–28 Feb 2006

Author Biographies

Professor Abdel Aziz Saleh was the Special Advisor to the Regional Director of the World Health Organization (WHO) Eastern Mediterranean Regional Office (EMRO) for Medicines. Before joining WHO, Professor Saleh was Professor of Pharmaceutics and Vice Dean for Students Affairs at the Faculty of Pharmacy, Alexandria University. He contributed to developing pharmacy education in Egypt, Nigeria and Libya. Since he joined WHO in 1988 as

Regional Advisor for WHO Essential Drugs Programme and then Deputy Regional Director, he was active in developing national drug policies. In consultation with regional experts, he developed regional guidelines for bioequivalence and stability studies. Professor Saleh provided technical support to national capacity building programmes in several pharmaceutical areas including drug quality assurance and post marketing surveillance systems. Recently, Professor Saleh supports countries of WHO Eastern Mediterranean Region to obtain WHO Prequalification status in Vaccines and Drugs Production.

Mrs. Kim Huynh-Ba is the Technical Director of Pharmalytik. She has over 24 years of experience in various analytical areas of pharmaceutical development and a primary focus in stability sciences, analytical development, outsourcing and technology transfer management. She has authored numerous technical publications and is invited frequently to present at national and international conferences. She is the editor of the "*Handbook of Stability Testing in Pharmaceutical Development: Regulations, Methodologies and Best Practices*", Springer.

Chapter 6

The Challenge of Diverse Climates: Adequate Stability Testing Conditions for India

Saranjit Singh, Amrit Paudel, Gaurav Bedse, Rhishikesh Thakare and Vijay Kumar

Abstract India has diverse climatic conditions across the country, which is the reason for interest in its long-term stability testing conditions that are not yet defined. The same were calculated using the risk-based approach proposed recently in the literature. In total, 18 cities were selected through length and breadth of the country, and mean kinetic temperature and average relative humidity were determined for each. The extreme climate was found in the city of Trivandrum in the southern state of Kerala, which has equatorial monsoonal conditions for almost 7 months in a year. The long-term stability testing condition for the place calculated out to be 30°C/70% RH. The same was considered to represent whole of the country, which being the worst case. Although this storage condition was originally suggested for zone IV, it is not included in current WHO classification of long-term stability testing conditions for various zones. This article describes a revised classification proposed to accommodate India and even other countries with hot and moderately humid climate.

1. India: The Country

India, the seventh largest country of the world, is South Asian region's largest sovereign and democratic republic. It is crowned by the world's highest mountain range "The Himalayas" and surrounded by Pakistan, Nepal, China, Bhutan and Bangladesh. It comprises of 28 States, 6 Union Territories and one National Capital Territory, Delhi.

It is the second most populated country, having a head count of ~1.2 billion. North to north-east belt in India is most populated. India's largest urban agglomerations are Mumbai, Kolkata, Delhi, Chennai, Bangalore, Hyderabad and Ahmedabad.

2. Pharmaceutical Industry in India

The pharmaceutical industry in India produces 80% of bulk drugs and entire range of formulations. There are 6,000 plus listed and unlisted companies,

From: *Pharmaceutical Stability Testing to Support Global Markets*: Pharmasp,
Edited by: K. Huynh-Ba, DOI 10.1007/978-1-4419-0889-6_6,

both multinational and local. The top 10 of them account for <40% of market, while >250 companies account for 70% share. The retail sales of pharmaceuticals in domestic market in 2006 were $8 billion, and as per estimates, Indian pharma market is set to grow to $20 billion by 2015. The Indian pharmaceutical industry is ranked fourth in the world, accounting for 8% of world's production by volume.

3. Diverse Physical and Climatic Conditions in India

Most of the India's northern states are situated in the Himalayan mountain range. The central and eastern India consists of the fertile Indo-Gangetic plain. In the west is the Thar desert. The southern Indian Peninsula is almost entirely composed of the Deccan plateau, which has two hilly coastal ranges, the western and the eastern ghats.

According to Köppen, which is an internationally established environmental system, India is truly diverse in its climate, hosting several major climatic subtypes: Alpine tundra and glaciers in the north, Arid desert in the west, Humid tropical regions supporting rainforests in the southwest and the island territories. Many regions have starkly different microclimates.

The nation has four seasons: Winter (January and February), Summer (March to May), Monsoon (rainy) (June to September), and Post-monsoon period (October to December).

4. Calculation of Long-Term Stability Test Conditions for India by Risk-Based Approach

Due to the intense focus on the country and owing to its diverse climate, it was considered prudent to work out appropriate long-term stability testing condition for India, using the risk-based approach proposed by Zahn to establish stability testing conditions in tropical countries (Zahn et al. 2006). The same protocol has been adopted by the World Health Organization (WHO) for classification of real-time stability test conditions for all its member states. The calculations for India had been pending.

The Zahn's procedure involves use of mean daily temperatures and dew points to calculate the daily and monthly fluctuations of temperature (T) and partial water vapour pressure (PD), the mean kinetic temperature (MKT) and the relative humidity (RH) (Zahn et al. 2006). Based on some collected data, the hottest and the most humid place in each country or region are identified to reflect the worst case. The safety margins for temperature (YT) and partial vapour pressure (YPD) are calculated taking into consideration the difference between measured meteorological parameters and the stability testing conditions. An appropriate long-term stability testing condition is proposed for the country or region, and on that basis it is classified into an appropriate climatic zone.

The climatic data for India were extracted from the ERA-40 re-analysis of European Centre for Medium-Range Weather Forecasts (ECMWF) (Uppala et al. 2005). Re-analysed screen level temperatures and dew points from 1979 to 2002, measured four times a day, were averaged into monthly means at four different coordinated universal times (UTCs) *viz.* 00UTC, 06UTC, 12UTC and 18UTC. Using the equations outlined above and the mean

Fig. 6-1. Selection of cities across India.

temperatures and dew points, MKT and RH were calculated for 18 populated cities covering whole of the country (Fig. 6-1). Eventually, the long-term storage conditions were derived along with calculation of risk-based safety margins YT and YPD.

Table 6-1 gives data for the 18 selected cities. It shows that Srinagar up in the north of India is the coldest among the selected cities, with mean temperature of 2.78°C, in line with alpine tundra conditions. Jodhpur in the west is the driest among all, as it falls in a region marked by Thar desert. Trivandrum and Chennai in the deep south have high temperature as well as high humidity, but between the two, the former has relatively lower temperature and higher humidity, whereas latter has slightly higher temperature but lower humidity. The data further show that YPD values were positive for all the cities at 30°C/65% RH, except Trivandrum, where positive YPD value was obtained only at 30°C/75% RH.

The 18 cities were then distributed according to WHO's climatic classification system (Table 6-2). The data showed that Srinagar was in Zone I, Jodhpur in Zone III, and all the other cities except Trivandrum fell in Zone IVA. Trivandrum alone fell in Zone IVB.

An exercise was subsequently carried out to repeat the calculations, as in Table 6-1, at 30°C/70% RH (Table 6-3), the original storage condition prescribed by WHO for Zone IV (WHO Technical Report Series 863,

Table 6-1. MKT, % RH, %YT and %YPD data for 18 cities across India at testing conditions of 30°C/65% RH and 30°C/75% RH.

Cities	T (°C)	MKT (°C)	YT (%)	PD (hPa)	RH (%)	RH % at 30 °C	Testing conditions (°C/% RH)	PD (hPa) at test conditions	YPD (%)
Srinagar	2.78	6.30	376.17	5.52	73.71	13.03	30/65	27.61	400.45
							30/75	31.85	477.31
Jodhpur	25.74	28.36	5.76	13.70	41.35	32.35	30/65	27.61	101.56
							30/75	31.85	132.51
Ahmedabad	26.87	28.07	6.88	19.38	74.59	45.76	30/65	27.61	42.46
							30/75	31.85	64.34
Mumbai	26.24	27.12	10.63	22.03	64.55	52.01	30/65	27.61	25.36
							30/75	31.85	44.61
Goa	25.58	26.33	13.94	21.86	66.62	51.62	30/65	27.61	26.29
							30/75	31.85	45.68
Trivandrum	27.31	27.48	9.17	29.14	80.20	68.80	30/65	27.61	-5.24
							30/75	31.85	9.31
Chennai	28.25	29.09	3.11	25.93	67.56	61.22	30/65	27.61	6.49
							30/75	31.85	22.85
Puri	26.69	27.37	9.62	26.16	74.69	61.78	30/65	27.61	5.54
							30/75	31.85	21.74
Kolkata	25.71	26.94	11.36	24.13	73.23	57.04	30/65	27.61	14.31
							30/75	31.85	31.86
Mizoram	22.40	23.35	28.51	20.11	74.41	47.49	30/65	27.61	37.27
							30/75	31.85	58.35
Cherapunji	22.42	23.50	27.67	21.56	79.71	50.92	30/65	27.61	28.05
							30/75	31.85	47.71
Patna	25.16	26.96	11.27	20.33	63.70	48.01	30/65	27.61	35.80
							30/75	31.85	35.80
Delhi	24.53	27.40	9.48	17.09	55.62	40.36	30/65	27.61	61.54
							30/75	31.85	86.35
Bhopal	25.71	27.78	7.98	15.76	47.80	37.22	30/65	27.61	75.16
							30/75	31.85	75.16
Nagpur	26.74	28.49	5.32	19.21	54.81	45.35	30/65	27.61	43.75
							30/75	31.85	65.83
Hyderabad	27.27	28.40	5.63	20.88	57.74	49.29	30/65	27.61	32.26
							30/75	31.85	52.57
Bangalore	24.99	25.71	16.68	20.99	66.46	49.56	30/65	27.61	31.55
							30/75	31.85	51.76
Amritsar	21.49	24.49	22.52	15.64	61.18	36.93	30/65	27.61	76.56
							30/75	31.85	103.67

Table 6-2. Assignment of climatic zone for selected cities of India according to current WHO classification criteria.

Cities	T (°C)	PD (hPa)	T/PD (°C/hPa)	RH % at 30°C	Storage condition (°C/% RH)	Climatic zone
Srinagar	2.78	5.52	≤15/≤11	13.03	21/45	I
Jodhpur	25.74	13.70	>22/≤15	32.35	30/35	III
Ahmedabad	26.87	19.38	>22/>15–27	45.76	30/65	IVA
Mumbai	26.24	22.03	>22/>15–27	52.01	30/65	IVA
Goa	25.58	21.86	>22/>15–27	51.62	30/65	IVA
Trivandrum	27.31	29.14	>22/>27	68.80	30/75	IVB
Chennai	28.25	25.93	>22/>15–27	61.22	30/65	IVA
Puri	26.69	26.16	>22/>15–27	61.78	30/65	IVA
Kolkata	25.71	24.15	>22/>15–27	57.04	30/65	IVA
Mizoram	22.40	20.11	>22/>15–27	47.49	30/65	IVA
Cherapunji	22.42	21.56	>22/>15–27	50.92	30/65	IVA
Patna	25.16	20.33	>22/>15–27	48.01	30/65	IVA
Delhi	24.53	17.09	>22/>15–27	40.36	30/65	IVA
Bhopal	25.71	15.76	>22/>15–27	37.22	30/65	IVA
Nagpur	26.74	19.21	>22/>15–27	45.35	30/65	IVA
Hyderabad	27.27	20.88	>22/>15–27	49.29	30/65	IVA
Bangalore	24.99	20.99	>22/>15–27	49.56	30/65	IVA
Amritsar	21.49	15.64	>22/>15–27	36.93	30/65	IVA

Table 6-3. MKT, %RH, %YT and %YPD data for 18 cities across India at testing condition of 30°C/70% RH.

Cities	T (°C)	MKT (°C)	YT (%)	PD (hPa)	RH (%)	RH % at 30 °C	Testing condition (°C/% RH)	PD (hPa) at test condition	YPD (%)
Srinagar	2.78	6.30	376.2	5.52	73.71	13.03	30/70	29.73	438.88
Jodhpur	25.74	28.36	5.76	13.70	41.35	32.35	30/70	29.73	117.04
Ahmedabad	26.87	28.07	6.88	19.38	74.59	45.76	30/70	29.73	53.40
Mumbai	26.24	27.12	10.63	22.03	64.55	52.01	30/70	29.73	34.98
Goa	25.58	26.33	13.94	21.86	66.62	51.62	30/70	29.73	35.99
Trivandrum	27.31	27.48	9.17	29.14	80.20	68.80	30/70	29.73	2.04
Chennai	28.25	29.09	3.11	25.93	67.56	61.22	30/70	29.73	14.67
Puri	26.69	27.37	9.62	26.16	74.69	61.78	30/70	29.73	13.64
Kolkata	25.71	26.94	11.36	24.15	73.25	57.04	30/70	29.73	23.08
Mizoram	22.40	23.35	28.51	20.11	74.41	47.49	30/70	29.73	47.81
Cherapunji	22.42	23.50	27.67	21.56	79.71	50.92	30/70	29.73	37.88
Patna	25.16	26.96	11.27	20.33	63.70	48.01	30/70	29.73	46.23
Delhi	24.53	27.40	9.48	17.09	55.62	40.36	30/70	29.73	73.94
Bhopal	25.71	27.78	7.98	15.76	47.80	37.22	30/70	29.73	88.61
Nagpur	26.74	28.49	5.32	19.21	54.81	45.35	30/70	29.73	54.79
Hyderabad	27.27	28.40	5.63	20.88	57.74	49.29	30/70	29.73	42.42
Bangalore	24.99	25.71	16.68	20.99	66.46	49.56	30/70	29.73	41.66
Amritsar	21.49	24.49	22.52	15.64	61.18	36.93	30/70	29.73	90.11

Annexure 5 (1996)). For Trivandrum, a positive value of YPD was obtained even for 30°C/70% RH, indicating that it was the true long-term condition for the city instead of 30°C/75% RH.

So an issue emerged pointing – what was the real stability test condition for India, for a country with a very diverse climate? Was it 30°C/65% RH considering an average for whole of the country, or 30°C/75% RH taking into account WHO's Zone IVB option, or rather 30°C/70% RH considering actual calculated value for Trivandrum?

The city of Trivandrum falls in Indian state of Kerala, which is the only state with equatorial monsoonal conditions for almost 7 months of a year and is hence the most stringent among all the others. Thus, it could not be ignored while determining the stability storage conditions for the country, as it was also one amongst the densely populated. Hence, 30°C/65% RH was ruled out as the condition for the country. Even the WHO's Zone IVB option (30°C/75% RH) was considered improbable, as no city or state in India including Kerala, matched the equatorial fully humid conditions, as in ASEAN or Brazil. Hence, the most justifiable and qualifying long-term stability test storage condition for India was determined to be 30°C/70% RH.

5. Suggested Revision of WHO's Classification of Global Climatic Zones

A suggestion emerged that current criteria used by WHO to classify climatic zones and testing conditions need to be modified to accommodate stability test condition for India, and for other countries with similar environmental distribution. The proposal is put forth in Table 6-4. An initial estimate indicated that there were more than 25 countries under WHO's umbrella, which had hot and moderately humid climate, similar to India. Hence, it is proposed that Zone IVA should cover hot and low humid climates with storage condition of 30°C/65% RH, a new zone IVB condition should be added to cover hot and moderately humid climates of 30°C/70% RH, while the originally proposed Zone IVB storage condition should be restricted to hot and very humid climates and converted as Zone IVC. The benefit of this change would be that which would allow countries with moderate environment to carry out stability tests under milder and appropriate conditions for with-in country sales of pharmaceutical products, thus avoiding under or over testing. It has been reported

Table 6-4. Proposed revision in WHO classification system.

Climatic zone	Definition	Criteria[a] (°C/hPa)	Testing conditions (°C/%RH)
I	Temperate climate	≤15/≤11	21/45
II	Subtropical and Mediterranean climate	>15–22/>11–18	25/60
III	Hot and dry climate	>22/≤15	30/35
IVA	Hot and low humid climate	>22/>15–27	30/65
IVB	Hot and moderately humid climate	>22/>27–30	30/70
IVC	Hot and very humid climate	>22/>30	30/75

[a]Mean annual temperature measured in the open air/Mean annual partial water vapor pressure

that drugs, both hygroscopic and non-hygroscopic, gain moisture rapidly in the region beyond 65–75% RH, which can be detrimental for moisture sensitive drugs (Visalakshi et al. 2005). Thus, long-term testing at 30°C/75% RH instead of 30°C/70% RH may mean exposing drug products to higher humidity than required, resulting in need of over-protection through packaging, leading in turn to increased costs. This typically applies to India too, where there are a large number of companies catering exclusively to 1.2 billion populace of the country. Of course, for companies with global presence, it may be a good sense to carry out testing in the harshest condition of 30°C/75% RH to cover sales to hot and very humid regions, like Brazil, ASEAN, etc.

6. Conclusion

India has a diverse climate comprising of all the zones, except Zone IVB, defined recently by WHO. In that respect, its climate is milder than hot and very humid regions, like Brazil, ASEAN, etc. The calculations indicate adequate long-term stability test storage condition for India as 30°C/70% RH, based on consideration of the harshest conditions prevalent in the well-populated state of Kerala in South India. This storage condition is not listed in the WHO's new classification, although it was originally prescribed by the agency for Zone IV. Hence it is suggested that storage condition for India and other countries with similar hot and moderately humid climate be accommodated in the WHO classification, by revising it through trifurcation of Zone IV into Zone IVA, Zone IVB and Zone IVC.

Acknowledgment The supply of ERA data and training in calculations by Dr Manuel Zahn are duly acknowledged.

References

Uppala SM, Kallberg PW, Simmons AJ, Andrae U, Da Costa Bechtold V et al (2005) The ERA-40 reanalysis. Q J R Meteorol Soc 131(612):2961–3012

Visalakshi NA, Mariappan TT, Bhutani H, Singh S (2005) Behavior of moisture gain and equilibrium moisture contents (EMC) of various drug substances and correlation with compendial information on hygroscopicity and loss on drying. Pharm Dev Technol 10:489–497

WHO, Guideline for stability testing of pharmaceutical products containing well established drug substances in conventional dosage forms. WHO Technical Report Series 863, Annexure 5 (1996)

Zahn M, Kallberg PW, Slappendel GM, Smeenge HM (2006) A risk-based approach to establish stability testing conditions for tropical countries. J Pharm Sci 95(5): 946–965 Erratum (2007) 96: 2177

Author Biography

Dr. Saranjit Singh is a Professor and Head of the Department of Pharmaceutical Analysis NIPER, SAS Nagar, India. He is leader of the stability testing and impurity profiling facility at the institute. Dr Singh is having interest in the area of stability testing for the last 26 years and has delivered more than 225 invited lectures. He has published 120 research papers, general articles and book chapters. He was the chairman of the committee constituted by Indian Drug Manufacturers Association (IDMA) for drafting stability test guidelines for Indian Industry. He is member, Editorial Advisory Board of Journal of Pharmaceutical and Biomedical Analysis (Netherlands). He is recipient of Professor M.L. Khorana Memorial Lecture Award, 2005 and IDMA-APA Outstanding Analyst Award, 2002.

Chapter 7

Requirements for South East Asian Markets

Lucky S. Slamet and Kim Huynh-Ba

Abstract ASEAN organization is composed of ten different countries, located in the region of Southeast Asia. The environmental conditions of these countries are mostly hot and humid, which would fall in the categories of Zone IV climatic condition. This region has issued a guideline and recommended a storage condition for stability studies of 30 ± 2°C with 75 ± 5% RH.

1. Background

Stability is an essential factor of drug quality, safety and efficacy. Alteration of medicines can appear to be with physical, chemical or microbiological attributes. With regards to the stability testing, the first purpose is *to provide evidence on how the quality of a drug varies with time under the influence of environmental factors such as temperature, humidity and light as to guarantee the product's safety*. The second purpose is to *establish a re-test period for the drug substance, a shelf life of the drug product and recommended storage condition*. The purpose of this paper is to share different perspectives on stability testing of pharmaceutical products for marketing in Southeast Asia.

2. Harmonizing Stability Requirements for ASEAN Zone IV

2.1. Consideration on Storage Condition of ASEAN Zone IV

The Association of South East Asian Nations (ASEAN) organization is composed of ten different countries, located in the region of Southeast Asia. In accordance to the first purpose of stability testing, a calculation based on meteorological data in ASEAN countries has demonstrated that long term stability condition or storage at 30°C/75% RH better reflects realistic condition in many ASEAN cities which are hot and humid. Storage condition at 30°C/65% RH may mean more risk for drug product with water mediated degradation, since the safety margin becomes much less or even non-existent considering the humidity condition in some parts of the region. Other consideration on

From: *Pharmaceutical Stability Testing to Support Global Markets*: Pharmasp,
Edited by: K. Huynh-Ba, DOI 10.1007/978-1-4419-0889-6_7,

the implementation of ASEAN Zone IV is that if calculations are done only for ASEAN cities which are generally more humid, different values will be obtained and the safety margin may become non-existent. Therefore, the average approach versus testing conditions towards more stressful rather than less stressful condition is needed to provide safety margin in favor of the patient (Zahn, 2008).

2.2. Harmonization Efforts

To harmonize the stability requirement, ASEAN Consultative Committee for Standard and Quality-Pharmaceutical Product Working Group (ACCSQ-PPWG) develops guidelines for stability standards and studies in ASEAN regions. The guidelines basically adopts ICH Q1A Guidelines with the exception of storage condition of real time studies, using storage condition of 30°C ±2°C/75% ± 5% RH.

3. ASEAN Guideline on Stability Study of Drug Product

3.1. Testing Frequency

Testing frequencies are similar to those suggested in Q1A (R2). It is summarized in Table 7-1.

3.2. Storage Conditions

Table 7-1 lists three storage conditions recommended by the organization: Real time, accelerated and alternatives to the accelerated study. Real time and accelerated data are recommended for New Chemical Entities (NCE), Generics and Variations (MaV and MiV). Alternatives to the accelerated study can also be used for Generics and Variations (MaV and MiV). Medicines are developed in different physical forms thus are required different storage conditions.

3.2.1. Solid Dosage Form

Stability studies of drug products in primary containers permeable to water vapor should be conducted in 30 ± 2°C/75 ± 5% RH. For drug products in primary containers impermeable to water vapor, studies should be done at in 30 ± 2°C without a specified relative humidity. Condition for accelerated studies is recommended to be done at 40 ± 2°C/75 ± 5% RH.

Table 7-1. Testing frequency.

Products	Storage condition	Testing frequency
NCE, generics and variations	Real time	0, 3, 6, 9, 12, 18, 24 months and annually through the proposed shelf-life
	Accelerated	0, 3 and 6 months
Generics and variations	Alternatives to the accelerated conditions	0, 1 and 3 months

Table 7-2. Conditions for new chemical entity drug products.

Study	Storage condition	Data needed
Real time	30±2°C/75±5% RH	12 months
Accelerated	40±2°C/75±5% RH	6 months

Table 7-3. Conditions for generics and variation.

Study	Storage condition	Data needed
Real time	30±2°C/75±5% RH	12 months
Accelerated	40±2°C/75±5% RH	6 months
Alternative accelerated	45–50°C and 75±5% RH	3 months

Table 7-4. Conditions for drug product stored in refrigerator.

Study	Storage condition	Data needed
Real time	5±3°C	12 months
Accelerated	25±2°C/60±5% RH	6 months

3.2.2. New Chemical Entity (NCE) Drug Products

Stability studies are done for at least three batches at the conditions listed in the following Table 7-2.

3.2.3. Generics and Variation (MaV and MiV if appropriate) Products

Similar to NCE products, stability studies are done at the same real time and accelerated conditions. This guideline is also composed of storage conditions of generics and variation with major and minor variations (MaV and Miv) products. In addition to the real time study and accelerated study, alternates to the accelerated study should also be considered for 3 months. For conventional dosage form and stable drug substance, two batches are required; however, for critical dosage form, a minimum of three batches would be needed (Table 7-3).

3.2.4. Drug Product Intended for Storage in a Refrigerator

A minimum of three batches are studied as given in Table 7-4.

3.2.5. Drug Product Intended for Storage in a Freezer

Real time study is to be done at –20±5°C for at least 12 months.

3.3. Container Closure System

Stability testing should be conducted on the drug packaged in the container closure system proposed for marketing. When moisture-permeable containers are considered, stability studies should be done under high humidity conditions. Specific parameters, such as the material's thickness and permeability coefficient, should be included when different permeability of various packaging materials is used.

3.4. Different Stability Study from ASEAN Guideline

The drug product for which the stability is conducted at lower temperature and humidity other than ASEAN specified conditions for real time studies, sufficient supporting data and technical basis must be provided. Real time studies are to be continued after approval to the end of expiry. The following post approval stability test can be conducted in any ASEAN member country, country of origin, or any country that can meet the required storage conditions.

3.5. Labeling

Statements or labeling should be based on the stability study. The storage conditions (temperature, light, humidity) should be labeled or refer to the recommendations below:

- Store below 30°C or do not store above 30°C (normal storage condition).
- Store below 25°C or do not store above 25°C (under controlled air-conditioning).
- Store between 2 and 8°C (under refrigeration, no freezing).
- Store below 8°C (under refrigeration).
- Store between −5 and 0°C (in freezer).
- Store below −18°C (in deep freezer).

The use of term such as "ambient conditions" or "room temperature" is avoided as it is misleading. General precautionary statements, such as "Protect from light" and/or "Store in a dry place," may be included; however, it should not be used to conceal stability problems. In addition, recommendations should also be made as to the utilization period and storage conditions after opening and dilution or reconstitution of solution. Specific requirements should be stated for drug products that cannot tolerate freezing.

4. ASEAN Position on Stability Studies in Global Environments

The WHO Stability Guideline divides the regions in ASEAN into two zones: Zone IVa and Zone IVb (Table 7-5).

Indonesia was mandated by the 12th ACCSQ – PPWG meeting and has conveyed to WHO-HQ a common position of ASEAN Member Countries that the ASEAN storage condition for stability study is Zone IVb, which is 30°C/75% RH. For established products in ASEAN region, there is no need to conduct study at new condition if there is no evidence of adverse events.

Table 7-5. Conditions for stability studies to support Zone IV.

Region	Storage condition
Zone IVa	30°C/65% RH
Zone IVb	30°C/75% RH

For product registrations that have been submitted prior to the implementation date, it would depend on the submitted countries since the date of implementation is only a formal start date.

If the real time data are supported only by 12 months data, extrapolation may be sought for up to twice the available real time data, but should not be more than 12 months beyond the period covered by real time data, depending on the change over time, variability of data observed, proposed storage condition and extent of statistical analyzes. Full testing per Finished Product Specification needs to be carried out for stability study using stability indicating testing.

5. Conclusion

As discussed, stability testing is a crucial component on drug development process. Although many guidelines have been developed, many issues continually raised, such as the requirement of storage condition on oral dosage form in climatic zone IV. Every issue has played a leading and important role to determine a model of stability studies in global environments as WHO guidelines in Zone 1–4. As mandated by ACCSQ-PPWG, Indonesia has conveyed to WHO that a common position of ASEAN Member countries on ASEAN storage condition for stability study is Zone IVb, namely 30±2°C with 75±5% RH.

Reference

Zahn M (2008) Global Stability Practices, in *Handbook of Stability Testing in Pharmaceutical Development: Regulations, Methodologies and Best Practices*, Huynh-Ba, Kim (Ed.), 43–92.

Author Biographies

Dr. Lucky S. Slamet obtained her degree in Pharmacy from the University of Indonesia and her master degree in Community Health and Epidemiology from Queen's University at Kingston, Canada. Her field of interest includes other regulatory matters of drug and biologicals, and Drug Information. Therefore, she had taken various training to enhance her regulatory performance, particularly in the field of Drug and Biologicals Policy Issue; Quality Assurance; Drug and Biological Evaluation for Marketing Approval; Patent Protection, TRIPS and Health etc. She also served as WHO Temporary Adviser on her personal capacity particularly in Technical meetings and Workshop on Vaccine and Biologicals.

Currently her post is Deputy for Therapeutic Product, Narcotics, Psychotropic and Addictive Substance Control, The National Agency of Drug and Food Control (NA–DFC), The Republic of Indonesia. Her responsibility includes pre-market evaluation on efficacy, safety and quality of drug and biologicals as well as post-market control of therapeutic products, narcotics, psychotropics and addictive substance such as enforcement of GMP, inspection on distribution channel, Adverse Drug Reaction Monitoring and Adverse Event Following Immunization Surveillance. Her professional activities include many public health programs, including those in collaboration with WHO and other similar organizations.

Mrs. Kim Huynh-Ba is the Technical Director of Pharmalytik. She has over 24 years of experience in various analytical areas of pharmaceutical development and a primary focus in stability sciences, analytical development, outsourcing and technology transfer management. She has authored numerous technical publications and is invited frequently to present at national and international conferences. She is the editor of the "*Handbook of Stability Testing in Pharmaceutical Development: Regulations, Methodologies and Best Practices*", Springer.

Chapter 8

The Role of USP Monographs in Stability Testing

Karen A. Russo

Abstract For drugs, including biologics, and excipients, the US Pharmacopeial Convention's (USP's) role in setting public standards generally begins after the Food and Drug Administration (FDA) approval. Working with pharmaceutical manufacturers, FDA, USP staff, and other stakeholders, USP's Council of Experts develops the public standard, i.e., a drug substance and drug monograph with allied reference materials (Official USP Reference Substances) when needed. The drug substance and product monographs contain tests, procedures, and acceptance criteria (collectively referred to as the specification) for an official article in the *United States Pharmacopeia* (*USP*) and excipients monographs are included in *National Formulary* (*NF*). Articles in *USP–NF* are expected to comply with the specification and other requirements until their expiration date in accordance with the adulteration and misbranding provisions of the US Food, Drug & Cosmetic Act, as well as in accordance to USP's requirements as stated in General Notices (see current *USP*). Other bodies, e.g., States and other countries may rely on standards in *USP–NF*. Because USP standards are applicable throughout the shelf-life of a product, they are important for assurance of adequate quality throughout the time that a medicine is available to the patients and consumers.

1. Introduction

1.1. United States Pharmacopeial Convention Facts: Who and What is USP?

Established by a group of volunteer physicians in 1820, USP is an independent nonprofit private organization and is also the only nongovernmental pharmacopeia worldwide (Anderson and Higby 1995). USP's headquarters are located in Rockville, Maryland, and there are international sites which are located in Basel, Switzerland; Hyderabad, India; Shanghai, China; and the newest site is in São Paolo, Brazil (opened in August 2008).

From: *Pharmaceutical Stability Testing to Support Global Markets*: Pharmasp,
Edited by: K. Huynh-Ba, DOI 10.1007/978-1-4419-0889-6_8,

1.2. USP Public Standards

USP's mission is to produce meaningful public standards by the public development of monographs that focus on the identity, strength, quality, and purity of official articles. The safety and efficacy of the products are the responsibilities of the Food and Drug Administration (FDA), and the data used to develop the monographs are generally provided by the pharmaceutical industry, including innovator and generic companies around the world[1]; (Bhattacharyya et al. 2004). Generic pharmaceutical companies are increasingly involved in the development of public standards. In recent years, USP received about the same number of new monograph submissions from generic companies as from innovator companies. USP–NF are named as official compendia of the United States in the Federal Food, Drug & Cosmetic Act in several ways, perhaps most notably in the adulteration and misbranding provisions of the Act. USP also provides compendia for foods, including dietary supplements (USP's dietary supplement section of USP) and food ingredients (Food Chemicals Codex). The provisions of these compendia are beyond the scope of this article.

2. Typical Monograph Procedures and Requirements

2.1. Definition with Proposed Limits

Each official article has a definition, and these definitions contain the proposed assay limits, typically 90.0%–110.0% of label claim for dosage forms and 98.0%–102.0% of label claim for drug substances. These values may vary depending on the contents of the monograph sponsor's application to FDA – for example in a New Drug Application (NDA) or an Abbreviated New Drug Application (ANDA) – and the contents of the submission to USP.

2.2. Identification Tests

Infrared (IR) spectroscopy is probably the most common identification technique used in the pharmaceutical industry. Other common procedures for identification include ultraviolet absorbance, chromatographic retention time, and measurement of counter-ions.

2.3. Packages and Labeling

The monograph's *Packaging and storage* statement describes the type of packaging, such as a tight container or blister packs, and the storage temperature recommended for the article. The *Labeling* statement in a dosage form monograph indicates the information needed on the product's package insert. In some cases, the monograph may define cautionary statements that must be included on the package and in the package insert.[2] For a drug substance, the labeling generally is the certificate of analysis.

---

[1] 21 USC 501.

[2] 21 USC 501(b).

2.4. Other Procedures and Requirements

Monograph submission packages presented by companies include the tests, procedures, and acceptance criteria (collectively known as the specification), and they reflect the private regulatory specification in an approved NDA or ANDA. Monographs include other procedures and requirements such as water determination, loss on drying tests, limits, and content procedures, along with dosage form-specific tests such as dissolution, viscosity, particulate matter, microbiological testing, and the test for *Enantiomeric purity* when appropriate.

3. USP Monographs

Most USP monographs include an assay for determining content, and ideally the assay is a stability-indicating procedure. Assays are intended to provide a means to determine the content and can be used as an indication of the stability of the article. Most compendial procedures for assays are based on the stability-indicating separation (e.g., chromatographic) procedures. Nonspecific procedures (e.g., titration) are also valid. The acceptance criteria should include manufacturing variability based on the FDA approval, the experimental error of the procedure, and sampling errors (Williams 2006; Hauck et al. 2008). Generally, the assay limits are between 90.0 and 110.0% for drug products and between 98.0% and 102.0% for drug substances. These ranges are widely accepted throughout the pharmaceutical industry, but individual monographs may indicate assay limits that differ from these traditional values.

Some assay procedures, such as a titration, may not be stability indicating alone but can be coupled with a stability-indicating procedure for the detection of impurities/degradants. Stability-indicating assay procedures can accurately and precisely determine contents in the drug substance or in the presence of excipients and other components (i.e., dosage forms).

The monograph as a whole may be considered stability indicating based on one or more tests and/or procedures that may include determination of water content, pH, or viscosity, which can be considered indicators of stability for a given official article.

4. Testing of Drug Products

The types of tests for dosage forms depend on the type of product being tested, but the main categories include solid oral, liquid, and parenteral dosage forms.

4.1. Solid Oral Dosage Forms

Table 8-1 lists typical testing recommended for solid dosage forms.

4.2. Liquid Oral Dosage Forms

Table 8-2 lists typical testing for liquid oral dosage forms.

4.3. Parenteral Dosage Forms

Table 8-3 lists typical testing performed on parenteral dosage forms.

Table 8-1. Tests for solid dosage forms.

• Dissolution/drug release/disintegration	• Uniformity of dosage units
• Friability	• Water content

Table 8-2. Tests for liquid oral dosage forms.

• Uniformity of dosage units	• Particle size distribution
• pH	• Redispersibility
• Microbial limits	• Rheological properties
• Antimicrobial preservative content	• Specific gravity
• Antioxidant preservative content	• Reconstitution time
• Extractables	• Water content
• Dissolution	• Alcohol content

Table 8-3. Tests for parenteral dosage forms.

• Uniformity of dosage units	• Extractables
• pH	• Functionality testing
• Sterility	• Osmolarity
• Endotoxins	• Particle size distribution
• Pyrogens	• Particulate matter
• Water content	• Redispersibility
• Antimicrobial preservative content	• Reconstitution time
• Antioxidant preservative content	

5. Labeling

Labeling statements related to safety may be included in USP monographs for dosage forms. Such labeling may include statements like "Dilute before use" and may feature tall man lettering to draw the user's attention to the label. Another use of labeling statements in monographs is to indicate which compendial tests and/or procedures are applicable to that official article. Statements such as these convey compendial requirements to the reader but are not related to the safety issues.

6. Identification

As previously noted, Infrared (IR) spectroscopy is the preferred technique for identification tests. For certain dosage form monographs, an IR procedure may be challenging because it requires isolation of the drug substance(s) from the sample matrix so that a spectrum can be obtained to ascertain the identity. Chromatographic procedures, such as chromatographic retention time or migration agreement, are commonly included as identification procedures. Ideally, a monograph includes at least two orthogonal identification procedures rather than a single procedure. If the dosage form contains more than one drug substance, then the monograph includes identification test(s) for each active component.

7. *USP–NF* General Chapters Related to Stability

Three main general chapters relate to stability: *Pharmaceutical Stability* <1150>; *Validation of Compendial Methods* <1225>; and *Impurities in Official Articles* <1086> (US Pharmacopeial Convention 2009a).

7.1. Pharmaceutical Stability <1150>

This chapter states that the monograph specifications of identity, strength, quality, and purity apply throughout the shelf life of the product. Stability of a dosage form refers to the chemical and physical integrity of the dosage unit and, when appropriate, the ability of the dosage unit to maintain protection against microbiological contamination. This chapter also includes definitions of the various temperature ranges such as controlled room temperature and the climatic zones used in international practice.

7.2. Validation of Compendial Methods <1225>

This chapter discusses the validation of compendial procedures and is in alignment with the relevant ICH guidelines. This chapter includes analytical performance characteristics typically included in the validation process, such as linearity, accuracy, precision, specificity, and others. These parameters are widely recognized by the pharmaceutical industry and by regulatory agencies. Validation parameters will vary depending on the type of procedure and whether the chapter offers guidance about which parameters should be evaluated based on the category of procedure. The chapter also indicates the type of validation information that should be included in submissions to USP.

7.3. Impurities in Official Articles <1086>

This chapter describes the compendial approaches and terminology related to impurity testing and discusses the comparisons between impurity testing in drug substances and dosage form monographs. In general, impurity tests are not repeated in the monographs for dosage forms when they are already included in monographs for bulk pharmaceutical chemicals and when the impurities are not expected to increase. This is a key piece of compendial information and facilitates the understanding of how impurities are addressed in the monographs.

According to this chapter, monographs for drug substances usually cite one of the three types of purity tests: (1) a chromatographic purity test coupled with a nonspecific assay; (2) a chromatographic purity-indicating method that serves as the assay; or (3) a specific test and limit for a known impurity, an approach that usually requires a reference standard for that impurity.

8. Impurities and Degradants

Procedures for the detection and quantitation of impurities and degradants frequently rely on the separation techniques such as high-performance liquid chromatography, gas chromatography, thin-layer chromatography, and other procedures. As discussed earlier, stability-indicating procedures are ideal.

Depending on the article, more than one test may be needed to identify the relevant impurities and/or degradants. Procedures submitted to USP should be validated in accordance with <1225> (US Pharmacopeial Convention 2007). Typical validation parameters for a stability-indicating quantitative impurity/degradant procedure include limit of detection, limit of quantitation, accuracy, precision, specificity, linearity, and forced degradation studies.

9. Degradants in Dosage Form Monographs

The degradants in dosage form monographs generally include specified and unspecified impurities as defined in ICH Q3B (R2) that may increase during the shelf life of the product (International Conference on Harmonization 2003). They are typically identified by the forced degradation stability studies conducted as part of the method development and validation process. This information is requested by USP to facilitate the monograph development or revision request processes.

Organic impurities are common, as evidenced by the impurities and degradants named in the monographs. For some articles, the relevant impurities and degradants cannot be detected using only a single technique or procedure. To address these situations, the monographs may include more than one test or procedure.

10. Flexible Monographs

Many single tests do not identify all formulation components, hydration states, polymorphs, or impurity profiles that result from various synthetic routes. A flexible monograph approach can successfully accommodate different analytical procedures and corresponding acceptance criteria. For drug substances, among the tests that may require flexible criteria are water, melting point, impurities, and possibly the assay. For many years, solid oral dosage form monographs have included multiple dissolution tests to accommodate formulations from various manufacturers. When a flexible approach is implemented in a monograph, a corresponding labeling requirement should indicate the test with which the article complies. Because some flexible tests involve different procedures, procedure-specific USP reference standards may be required for testing.

11. Examples of Flexible Monographs

11.1. Loratadine Drug Substance

This is an example of a drug substance monograph that has two impurity tests to accommodate different routes of synthesis. In this example, *Test 1* is the default; articles using *Test 2* need to indicate this on the labeling. For drug substances, the certificate of analysis typically is considered the labeling.

Labeling –

If a test for *Related compounds* other than *Test 1* is used, then the labeling states with which *Related compounds* test the article complies.

Related compounds –

NOTE – On the basis of the synthetic route, perform either *Test 1* or *Test 2*. *Test 2* is recommended if 4,8-dichloro-6,11-dihydro-5*H*-benzo[5,6] cyclohepta[1,2-*b*]pyridin-11-one is a potential related compound.

TEST 1 – …

TEST 2 – …

11.2. Theophylline Extended-Release Capsules

This is an example of a solid oral dosage form monograph that includes 10 dissolution tests to accommodate various manufacturers. The labeling requirement indicates that each product must specify the dissolution test with which the article complies:

Labeling –

The labeling indicates whether the product is intended for dosing every 12 or 24 h, and states with which in vitro *Dissolution Test* the product complies.

Dissolution <711>

TEST 1 – …

TEST 2 – …

…

TEST 10 – …

11.3. Other Examples

Other official monographs use the flexible approach for various tests:

- Ethinyl estradiol drug substance.
- Meloxicam drug substance.
- Paroxetine drug substance.
- Potassium chloride extended-release tablets.

12. USP Standards-Setting Process

Requests to add a new monograph or revise an existing monograph follow the USP standards-setting process (US Pharmacopeial Convention 2007). In the small-molecules area, monographs usually follow one of the two main paths for submission: publication in *Pharmacopeial Forum* (*PF*) for submissions from sponsors who have received FDA approval; and the Pending Monographs approach for sponsors who have not yet received FDA approval of their application. A third pathway, Non-US Monographs, is available for drugs that treat neglected infectious diseases and are legally marketed outside the US.

12.1. *USP–NF* Monographs

Monograph development begins when USP receives a submission from a sponsor, typically a pharmaceutical company or drug substance manufacturer (Figure 8-1). Each monograph is assigned to a Scientific Liaison who reviews the data and information submitted by the sponsor and prepares a draft of the revision proposal, working with the sponsor to resolve any issues. The Scientific Liaison also collaborates with the assigned USP Expert

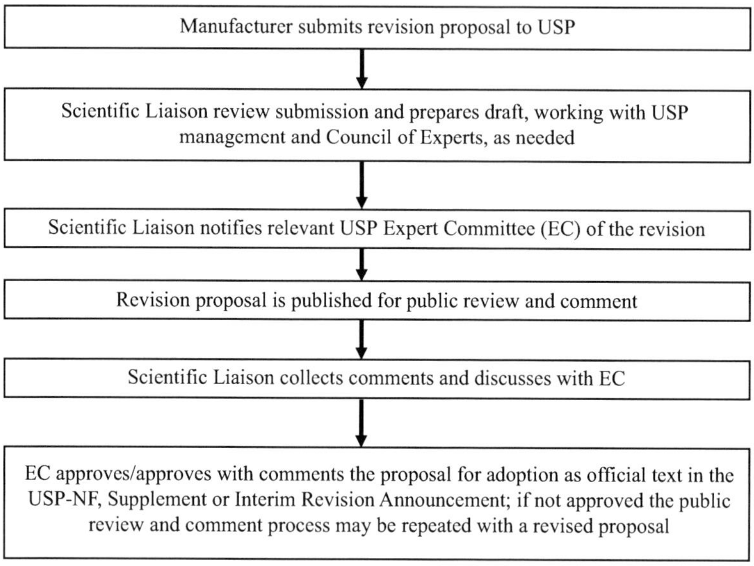

Fig. 8-1. The USP monograph development and revision process.

Committee(s) regarding the request for revision. Expert Committees are composed of elected subject-matter experts who, operating under strict conflict of interest rules, volunteer to assist with the development of monographs and general chapters in *USP–NF* (Schniepp 2006). The draft proposal is published in *PF* for a 90-day public review and comment period. If any comments are submitted during this period of public review, the Scientific Liaison shares them with the Expert Committees.

The relevant Expert Committees vote during a balloting process after the public review and comment period ends. If the draft monograph is approved for official adoption by the Expert Committees then the proposal is ultimately published in *USP–NF* or one of its two annual *Supplements*. The official text is the public standard and is legally enforceable by FDA. If the proposal is not approved for adoption, it may require revision and another round of publication in *PF*, public review and comment, and balloting by the relevant Expert Committees.

12.2. Pending Monographs

The Pending Monograph process was developed for manufacturers who are in the process of submitting or in the process of obtaining FDA approval for their ANDA (US Pharmacopeial Convention 2008). The process for submitting a Pending Monograph is nearly identical to the traditional path: The Pending Monograph path differs because the draft proposals are published on the USP Web site, rather than in *PF*, for the 90-day public notice and comment period. Following a review of all the comments received, the assigned Expert Committee(s) vote on the proposal. If approved, the draft proposal advances to the status of "Authorized" as opposed to "Official" – only the latter appear in *USP–NF* and are enforceable by FDA (see Footnote 2). The Authorized Pending Monograph remains on the USP Web site until the sponsor obtains

FDA's approval of the sponsor's ANDA (or, in the case of a drug substance, an ANDA citing the applicant's drug master file). After the sponsor obtains FDA approval, the Pending Monograph is eligible for adoption as official text in *USP–NF* or one of its *Supplements*.

Launched in February 2007, the Pending Monographs initiative enables monograph development to begin prior to FDA approval with the goal of establishing the public standard as quickly as possible after FDA approval of the monograph sponsor's application.

12.3. Non-US Monographs

Consistent with USP's mission "To improve the health of people around the world through public standards and related programs that help ensure the quality, safety, and benefit of medicines and foods," the Non-US Monographs initiative creates public standards for drugs and their dosage forms that are legally marketed outside the US (US Pharmacopeial Convention 2009b). The goal of these efforts is to support testing by first (e.g., manufacturers), second (e.g., payors), and third (e.g., government control laboratories) parties. Non-US Monographs represent a relatively narrow category of monographs for drugs for neglected infectious diseases (HIV-AIDS, tuberculosis, malaria, and others); public standards for such drugs are in high demand in regions outside the US. Monographs in this category follow a process like that of the other monographs with one major difference: Non-US Monographs are published only on the USP Web site and are not intended to advance into *USP–NF*.

13. Summary

USP monograph specifications apply throughout the shelf life of the official article. Although some monograph tests may be on their own stability indicating, the monograph as a whole is intended to be stability indicating, which may require more than one test to properly assess the article. Aspects of stability testing are addressed in various *USP–NF* general chapters.

References

Anderson L, Higby GJ (1995) The spirit of voluntarism: a legacy of commitment and contribution – the United States Pharmacopeia 1820–1995. US Pharmacopeial Convention, Rockville, MD

Bhattacharyya L, Cecil T, Dabbah R, Roll D, Schuber S, Sheinin EB, Williams RL, and the USP Council of Experts Executive Committee (2004) The value of USP public standards for therapeutic products. Pharm Res 21:1725–1731. doi:10.1023/B:PHAM.0000045222.01170.4c

Hauck WW, Koch W, Abernethy DR, Williams RL (2008) Making sense of trueness, precision, accuracy, and uncertainty. Pharm Forum 34:838–842

International Conference on Harmonization (2003) Q3B (R2) impurities in drug products. http://www.ich.org/cache/compo/276-254-1.html. Accessed 22 Apr, 2008

US Pharmacopeial Convention (2007) USP guideline for submitting requests for revision to USP–NF. www.usp.org/pdf/EN/USPNF/revisionGuide.pdf. Accessed 22 Apr 2008

US Pharmacopeial Convention (2008) US Pharmacopeia pending monographs guideline. www.usp.org/standards/pending/faq.html. Accessed 22 Apr, 2008

US Pharmacopeial Convention (2009a) USP 32–NF 27, Pharmaceutical Stability <1150>, Validation of Compendial Methods <1225>, and Impurities in Official Articles <1086>. US Pharmacopeial Convention, Rockville, MD, pp. 662–663, 733–736, 546–549

US Pharmacopeial Convention (2009b) USP non-US monographs guideline. www.usp.org/standards/international/guidelines.html. Accessed 22 Apr, 2008

Schniepp S (2006) Understanding the United States Pharmacopeia and the National Formulary: demystifying the standards-setting process. Parenteral Drug Association, Bethesda, MD

Williams RL, Project Team 4, the 2000–2005 Reference Standards Committee of the USP Council of Experts and Its Advisory Panel, and USP Staff and Consultant (2006) . Official USP reference standards: metrology concepts, overview, and scientific issues and opportunities. J Pharm Biomed Anal 40:3–15. doi:10.1016/j.jpba.2005.07.017.

Author Biography

Dr. Karen Russo is Vice President, Small Molecules at USP. In her current position, she leads a group of scientists who are responsible for approximately 3000 drug substance and dosage form monographs in *USP–NF*. Activities include the development of new monographs and revision of existing monographs. Dr. Russo joined USP in 2002 as a scientific liaison. Prior to joining USP, she worked as a clinical research associate monitoring human clinical trials and at a contract laboratory. Dr. Russo received a B.S. in Chemistry and Biology from Murray State University in Murray, Kentucky, and a doctorate in Pharmacology/Pharmaceutical Sciences from the University of Missouri at Kansas City.

Chapter 9

Regulatory Requirements for Stability Testing of Generics

Gary Buehler and Kim Huynh-Ba

Abstract The Office of Generic Drugs (OGD) of the Food and Drug Administration (FDA) believes that stability of drug products is best ensured through quality-by-design. Quality-by-Design (QbD) in the context of stability means that a sponsor identifies potential mechanisms for instability and designs the formulation, container-closure system and manufacturing processes to mitigate the potential for instability. Under OGD's new question-based review system, quality-by-design is documented in the pharmaceutical development report. However, the stability of generic drug products must still be demonstrated with data. On June 1, 2006, the FDA withdrew the 1998 draft stability guidance leaving the ICH Q1 documents as the only advice to sponsors. The scope of the ICH Stability Guidance is limited to new molecular entities and associated products. However, many sections of the ICH Q1 documents could apply to the new Abbreviated New Drug Applications (ANDAs). For example the storage temperature and relative humidity that are specified in ICH Q1 for accelerated, long term, and other conditions could be used for ANDAs. This talk will attempt to cover those recommendations which differ from ICH Q1 (e.g., initial submissions for new ANDAs), those for which no guidance is provided in the ICH document (e.g., post-approval changes), and those for which additional information is provided (e.g., data to support in-use conditions).

1. Introduction

Stability data demonstrates the product quality and preservation of therapeutic benefit over the product shelf-life. The stability profile must be established during drug development. Consumers expect stability of the drug product, as well as they expect that the Food and Drug Administration (FDA) to assure the drug product is stable over its labeled shelf-life. The Office of Generic Drugs (OGD) believes that the stability of drug products is best insured through Quality-by-Design (QbD). To simplify the process, the OGD recommends a Question-based-Review (QbR) submission.

From: *Pharmaceutical Stability Testing to Support Global Markets*: Pharmasp,
Edited by: K. Huynh-Ba, DOI 10.1007/978-1-4419-0889-6_9,

QbR is a general framework of submitting an Abbreviated New Drug Application (ANDA). It contains important scientific and regulatory review questions to comprehensively assess critical formulation and manufacturing process variables in order to set regulatory specifications, and to determine the risk level associated with the product design. Under QbD, a sponsor identifies the potential mechanisms of instability and designs formulations, manufacturing processes, and selects container/closure systems to mitigate the potential risk associated with stability.

The stability-related QBR questions were emphasized in the following sections.

2. Intrinsic Stability of a Drug Substance

Stability of the drug substance is very important. Questions related to its stability should be discussed. Stress testing and identification of appropriate degradation products should be included. Companies should understand the instability pathways of the active pharmaceutical ingredient (API).

3. Compatibility

Evidence should be included to demonstrate the compatibility between excipients and the drug substance. Although many generics use a set of universal excipients, the activity of these excipients in the proposed formulation may not be well understood. In addition, excipients may be reactive, which may lead to unintended consequences or interactions with a certain class of API.

As an example, in one oral solution, saccharin interacted with the active ingredient and caused an unacceptable product. Another application showed its API interacted with lactose to hasten degradation. Another case showed that the API interacted with triacetin used in the film coating of the tablet.

It is recommended that sponsors provide literature information on drug compatibility, if available. Companies should demonstrate that the excipients selected are suitable for the proposed formulation. Excipients could also greatly influence the drug substance's stability, safety, manufacturability and performance. Data from compatibility studies should also be submitted including studies under stress conditions. Information on trial formulations exposed to stress conditions is helpful. A simple visual inspection of mixed components for drug excipient compatibility is not adequate.

4. Container Closure Attributes

For generic products, specific information of container closure attributes is necessary to ensure product performance and stability of drug products. Factors affecting drug product quality such as special storage, light resistant containers, moisture protection, and an inert atmosphere should be understood during the development and included in the quality overall summary. Companies must demonstrate the suitability of the proposed container closure system. Extractables, leachables, dye from labeling, and other similar factors should be considered.

Examples of problems which have occurred with container/closure systems for an injectable antibiotic include:

- Sub-potency resulting from absorption of the active ingredient by the container rubber stopper.
- Super-potency caused by evaporation losses across the container/closure system.
- Particulates from the extraction of calcium stearate from stopper due to high pH of drug product.

Other questions related to the Drug Product attributes and manufacturing process are included in QbR to provide information in the submission package:

- What attributes should the drug product possess?
- How is the drug product designed to have these attributes?
- Why was the manufacturing process selected for this drug product?
- How are the manufacturing steps or the unit operations related to the drug product quality?
- Why the current formulation was selected?
- How the manufacturing process was connected?

An example showed that processing of antibiotic powder for oral suspension in ambient conditions with uncontrolled humidity had an adverse effect upon the shelf life of the product.

5. Expedite ANDA Submission Review

It is required that a new drug application (NDA) contain data on three batches with 6 months accelerated stability data. Expiry could then be extrapolated to 6 months past the available room temperature data. However, an ANDA has the requirement for only one batch with 3 months accelerated data to support a 24-months expiry. The difference in the amount of data is required because the NDA is for new drug substance, for which available data is limited. In an ANDA, there is a large body of stability data available, especially from the reference product. The average time for ANDA approval is approximately 18 months.

Since the Agency routinely receives multiple applications for the same product, the QbR format will guide the reviewers to prepare a consistent and comprehensive evaluation of the ANDA. It is noted that by July 2007, 90% of ANDA submissions were successfully done using the QbR process. Under the QbR system, the sponsor would present a summary of the application with data, which aid the reviewer in the application evaluation. This data set is used to justify the drug product shelf life as well as the storage conditions. Stability data at submission is based on the data available and the understanding of the product.

6. Regulatory Requirement for Generic Submission

Application submitted based on QbR concept does not indicate that stability testing is no longer necessary under QbD. The FDA continues to require stability data to verify the quality of the product; however, it is still not clear how much stability data would be necessary. This would depend greatly on the history of available data as well as extrapolation to long term data. Therefore, at the time of application, a stability package of the to-be marketed product combined with the knowledge gained during the product development allows a proposal of a reasonable tentative shelf-life of the drug product. Quality-by-design would help in the calculation of an expiry date and the stability of the drug product of interest.

In June 2006, both the 1987 stability guidance and the 1998 draft revision of that guidance were withdrawn. ICH Q1 documents are available for sponsors but the scope of ICH Stability Guidance is limited to new molecular entities, with some exceptions, and other associated products. The draft guidance, though withdrawn, still provides reasonable recommendations. In addition, *many sections of the ICH Q1 also apply to ANDAs such as storage temperatures, relative humidity, specified in ICH Q1 for accelerated long term and other conditions could be used for an ANDA.*

6.1. Stability Data

The only difference would be the number of batches and the amount of data needed for submission. An ANDA requires one batch at long term and 3 months accelerated stability data (1, 2 and 3 months time points are measured) with the attendant room temperature stability that will be accrued during the review period (Table 9-1). Similar requirements are also recommended for the Drug Product (Table 9-2). Data for in-use conditions will be requested for reconstituted products.

6.2. Batch Scale

Drug substance and drug product batches should be at least pilot scale, which is at least one-tenth of a full production scale. They must be manufactured by a procedure fully representative of and simulating that to be utilized in a full production scale batch.

Table 9-1. Recommended ANDA stability package: drug substance.

Study	Storage condition	Minimum time period covered by data at submission	Number of batches
Long-term	25±2°C 60±5% RH	Provide additional data in amendments	One batch
Accelerated	40±2°C 75±5% RH	3 months (report data at 0, 1, 2, and 3 months)	One batch

Table 9-2. Recommended ANDA stability package: drug product.

Study	Storage condition	Minimum time period covered by data at submission	Number of batches
Long-term	25±2°C 60±5% RH	Provide additional data in amendments as it becomes available	One batch
Accelerated	40±2°C 75±5% RH	3 months (report data at 0, 1, 2, and 3 months)	One batch

For some drug products, three batches are required for stability studies. These exceptions to the one batch requirement include certain complex drug products prepared from a critical intermediate that would use extended release or modified release (MR) beads or a transdermal system laminate. Three different lots of the critical intermediate should have been used in these three batches.

Nasal suspensions and inhalation drug products require three batches at the time of submission. A three batch requirement is being considered for all MR dosage forms, liposomal formulations and other novel delivery systems.

7. Frequent Generic Related Issues

7.1. Extrapolation

Routinely, a satisfactory 3 months accelerated data submission may permit 24-months tentative expiry providing that the room temperature data also meet specifications. If failure is observed at accelerated conditions, data of intermediate condition will be taken into consideration. If no intermediate condition data is available, then the exact length of the tentative expiration date would be determined on a case-by-case basis.

7.2. Labeling

The OGD recommends USP's labeling storage statements disregarding the label requirement from older references. If the drug product meets the appropriate acceptance criteria when tested after storage at the appropriate Q1A recommended storage conditions, then its label should support USP labeling statements. In the event that the reference product labeling statement is more restrictive than USP-controlled room temperature storage and the generic drug is demonstrated to be stable at controlled room temperature as well as refrigerated storage, the generic drug may be labeled as Room Temperature and/or Refrigerated condition.

However, if the reference product is labeled for USP controlled room temperature, but the generic version needs refrigerated conditions, OGD would generally not approve the generic application as this difference may cause confusion to the pharmacies.

8. Conclusion

In conclusion, the OGD encourages sponsors to fully address the QbR stability questions. These questions form the foundation to create a Quality-by-Design application. The feedback from the generic industry is that the QbR process stimulates discussion among development teams. Under Quality-by-Design principles, the Agency believes that the stability data are still needed to verify the product quality and to establish a shelf life consistent with that quality.

Author Biographies

Mr. Gary Buehler was appointed Director, Office of Generic Drugs, Center for Drug Evaluation and Research, in July of 2001. He had held the position of Deputy Director, Office of Generic Drugs, since May of 1999. Mr. Buehler has worked for FDA since 1986. Prior to joining the Office of Generic Drugs, he was a Senior Regulatory Project Manager in the Division of Cardio-Renal Drug Products in the Office of New Drugs. Mr. Buehler retired from the United States Public Health Service (USPHS) in April of 2000. He served at a variety of duty stations in the USPHS Division of Hospitals and Clinics as well as serving in the Indian Health Service in Nevada and Montana before coming to FDA. He graduated from Temple University School of Pharmacy.

Mrs. Kim Huynh-Ba is the Technical Director of Pharmalytik. She has over 24 years of experience in various analytical areas of pharmaceutical development and a primary focus in stability sciences, analytical development, outsourcing and technology transfer management. She has authored numerous technical publications and is invited frequently to present at national and international conferences. She is the editor of the "*Handbook of Stability Testing in Pharmaceutical Development: Regulations, Methodologies and Best Practices*", Springer.

Chapter 10

Stability Design for Consumer Healthcare Products

Jeffrey T. Needels, Mary W. Seibel, Karen L. Lucas, and Rachael Carlisle Roehrig

Abstract In the August of 2002, the Consumer Healthcare Products Association (CHPA) member companies were asked to assist the healthcare industry in addressing the lack of stability guidelines that were directly applicable to Over-the-Counter (OTC) monograph drug products. A Stability Working Group (SWG) was formed in May 2003 with experts from the following companies: Johnson & Johnson, Colgate-Palmolive Company, Novartis Consumer Health, Inc., McNeil Consumer Healthcare, Schering- Plough HealthCare Products, Inc., Bayer Health-Care, LLC, Purdue Pharma, The Procter & Gamble Company, Perrigo Company, and a CHPA facilitator. The SWG set out to develop a "white paper" document that would provide a science-based, industry recommended, best practice and minimum requirements for the development of OTC monograph drug products in the United States that are not regulated by an NDA/ANDA.

1. Stability Requirements for OTC Drug Products

What stability testing should be required for OTC drug products? To answer this question the SWG looked at the requirements for new drug products that are regulated by an NDA/ANDA (e.g., drug products requiring a prescription). See the International Conference on Harmonization (ICH) series of stability guidelines, http://www.ich.org/cache/compo/276-254-1.html. For these drug products containing new chemical entities (NCEs), the parent stability guideline ICH Q1A (R2), "Stability Testing of New Drug Substances and Products," requires three stability batches. The registration file must contain full testing for samples from each batch stored through 6 months at accelerated storage conditions and through 12 months at the long-term storage condition. In addition, the testing at the long-term storage condition is to continue throughout the proposed shelf-life period. Sufficient safety data must be available in order to gain approval for new drug products containing NCEs. Since extensive safety data through broad usage of the product is typically not available, it makes sense to require full stability profiling for registration. However, this is most likely too conservative for OTC Monograph drug products where the

From: *Pharmaceutical Stability Testing to Support Global Markets*: Pharmasp,
Edited by: K. Huynh-Ba, DOI 10.1007/978-1-4419-0889-6_10,

active drug substance is typically well characterized and has been shown to have an established safety profile. In point of fact, the SWG member companies do not test their OTC drug products to the level required for drug products containing NCEs.

The stability requirements for generic drug products are also well known. Typically, by the time a generic drug product is commercialized, the innovator's product has been on the market for some time and the experience and safety profiles become better known and understood. The FDA requires less stability data for generic drug products than for new chemical entities. One batch instead of three is required and stability testing must be completed through 3 months at accelerated storage conditions for inclusion into the registration file. A commitment to continue testing at the long-term storage condition throughout the proposed shelf-life period is also made. Generic drug products, as with drug products containing NCEs, require a prescription and the need for patients taking these products to be monitored with oversight by a physician even though generic drugs require less stability data than the original NCE product as the original product has been in the market for a period of time prior to going generic.

In looking at the stability requirements for new (NCE) and generic drug products and given the added commercial market experience and safety profile required for a drug to be included in the OTC monograph, it is scientifically sound to presume that the formal stability requirements for OTC monograph products should not be more conservative than those for either new or generic drug products. In September 2004, the first OTC stability guideline was written by the SWG member companies. The white paper document establishes minimum standards for the stability requirements of the newly developed OTC drug products: The document is entitled, "Guideline for the Stability Testing of Non-Prescription (OTC) Drug Products not regulated by an NDA/ANDA" and is available at http://www.chpa-info.org/scienceregulatory/Voluntary_Codes.aspx.

Table 10-1 provides a summary of the key differences between the stability requirements recommended by the SWG group for OTC drug products and the more conservative ICH Q1A (R2) stability requirements for new chemical entities.

Table 10-1. Minimum requirements for development stability (comparison between ICH Q1A(R2) and SWG guidelines).

	ICH Q1A(R2) *(new chemical entities)*	**SWG guideline** *(OTC drug products)*
# Batches	Three	At least 1
Batch size	Two 1/10th scale batches 3rd batch can be of a smaller scale	One 1/10th scale (a smaller batch size may be justified scientifically)
Testing time points	Long term: 0, 3, 6, 9, 12, 18, 24, and 36 month test intervals Accelerated: minimum of 4 test intervals (e.g. 0, 1, 3, and 6)	Long term: 0, 3, 6, 12, 24, and 36 month test intervals Accelerated: minimum of 3 test intervals (e.g. 0, 1, and 3)
Photostability	At least 1 Batch	May be omitted with Scientific Justification
Expiry dating	6 months accelerated data 12 months long-term data	3 months accelerated data (possibly less with body of data)

2. Stability Requirements for Changes to Currently Marketed OTC Drug Products

After working on the parent OTC stability guideline, the SWG began to address how changes to the currently marketed OTC drug products might be different, if at all, from changes required of drug products governed by new drug applications. What stability is required in order to make a change to a currently marketed OTC drug product? This is a question that requires thought and experience from stability scientists as OTC companies make a number of changes to products that are marketed commercially. There are many different types of changes and each can require a different answer to the question about what stability is required. The SWG used the Scale-Up and Post Approval Changes (SUPAC) document for immediate release of drug products as a baseline to develop a second OTC stability guideline in June of 2006. See SUPAC-Immediate Release guideline on the FDA website: http://www.fda.gov/. The SWG OTC stability Changes guideline has recently entered the final stages of development in October 2008. The document is entitled, "Voluntary Guideline for the Stability Testing in Support of Changes to Non-Prescription (OTC) Drug Products not regulated by an NDA/ANDA," and will be referred to as the SWG Changes Document hereafter. This guideline is available at http://www.chpa-info.org/scienceregulatory/Voluntary_Codes.aspx.

3. Classification of Changes for OTC Drug Products

The SWG Changes Document classifies many different types of changes for OTC monograph drug products as either a minor, moderate, or major change depending on the potential for significant affects to be observed in the stability profile of the drug product. The types of changes outlined in the guideline include: changes in manufacturing, packaging, or testing site; changes in formulation, process, or equipment; changes in batch size; and changes in primary packaging configuration. Based on the change classification of minor, moderate, or major, the document specifies whether stability data should be collected "prior to" or "concurrent with" commercialization of the changed product, or whether any stability is even required. Table 10-2 provides a summary of these classifications and the associated stability requirements for OTC drug products.

Table 10-3 lays out the differences in stability requirements used by industry for the changes made to OTC monograph drug products, specifically manufacturing site changes, as compared to those required by the SUPAC document for drug products governed by a new drug application.

Changes made to container/closure or primary packaging configurations were not addressed in the SUPAC guideline. However, these types of changes are addressed in the SWG Changes Document. Table 10-4 provides a sampling of the extent of stability requirements that might be required "prior to" or "concurrent with" the commercialization of a change proposed for the container/closure system for an OTC drug product.

The reader is referred to the actual guideline referenced above for a complete list of all of the OTC stability requirements for changes that might be proposed

Table 10-2. Stability data packages to support differing scenarios of product change.

Type of change	Pre-market Stability data	Post-market Stability data	Expiry period
Minor	None	None beyond regular annual batches	Maintain current expiry if supported; can market product immediately
Moderate	Comparative accelerated data on 1 batch; optional based on product history/knowledge base	1st production batch (minimum of one batch) on long-term stability through expiry period	Maintain current expiry if supported; can market product immediately
Major	3 months comparative accelerated and long term data generated up-front on minimum one batch of drug product with the proposed change	1st production batch (minimum of one batch) on long-term stability through expiry period	Maintain current expiry if supported; market product after 3 months comparative data

Table 10-3. Comparison of SUPAC vs. OTC guidance for manufacturing site changes.

Type of Change	Pre-Market Stability Data	Post-Market Stability Data	Expiry Period
Level 1 Changes Manufacturing site changes within a facility with the same equipment, SOPs, environmental conditions, controls, personnel (e.g. Remodeling an existing building or adding on to the existing facility)			
SUPAC Level 1 (Minor)	None	No extra data beyond regular annual batches	File data in annual report Maintain current expiry
OTC: Minor	None	No extra data beyond regular annual batches	Maintain current expiry; market immediately.
Level 2 Changes Change within a contiguous campus, or between facilities in adjacent city blocks, with the same equipment, SOPs, environmental conditions, controls, personnel			
SUPAC Level 2 (Moderate)	None	1st batch long term	CBE, file data in annual report Maintain current expiry
OTC: Minor	None	No extra data beyond regular annual batches	Maintain current expiry; market immediately.
Level 3 Changes with Available Significant Body of Information (SBI) Manufacturing site changes to a different facility/campus with the same equipment, SOPs, environmental conditions, controls			
SUPAC Level 3 (Major)	1 batch 3 months accelerated	1st production batch long term	Prior approval supplement
OTC: Moderate	None	1st production batch long term	Maintain current expiry; market immediately.
Level 3 Changes without Significant Body of Information (SBI) Manufacturing site changes to a different facility/campus with the same equipment, SOPs, environmental conditions, controls			
SUPAC Level 3, (Major)	Up to 3 batches, 3 months accelerated	1st (up to 3) production batches long term	Prior approval supplement
OTC: Moderate	1 batch 3 months accelerated	1st production batch long term	Maintain current expiry; market immediately.

Table 10-4. Container/closure changes.

Type of change	Pre-market Stability data	Post-market Stability data	Expiry period
Closure changes with inner seal unchanged (Minor)	None	No extra data beyond regular annual batches	Maintain current expiry; market immediately
Change to secondary packaging (Minor)	None	No extra data beyond regular annual batches	Maintain current expiry; market immediately
Change size of container/ closure within approved range (Minor)	None	No extra data beyond regular annual batches	Maintain current expiry; market immediately
Change size of container/closure outside of approved range (Moderate)	With SBI[a]: None Without SBI[a]: 1 batch, 3 months accelerated	1st batch long term	Maintain current expiry; market immediately
Change to different container/closure (Major)	Without SBI[a]: 1 batch, 3 months accelerated	1st batch long term	Maintain current expiry if supported; market product after 3 months comparative data

[a]Significant body of information (SBI)

to OTC drug products. As usual, these are general recommendations and basic industry practices taking into consideration the overall understanding that the OTC drug product is already well characterized and has an established safety profile. It is recommended that each particular situation is evaluated taking into account the known scientific information and existing stability profile data of the specific product and proposed change. This becomes especially true if more than one change is requested as this potentially could compound the degree of risk that might be expected and warrants additional caution.

4. Conclusion

The SWG continues to meet and address additional stability issues that are not well understood by the consumer healthcare industry. The intent is to provide a science-based risk management approach that can aid the industry in outlining minimum requirements in common practice by other consumer healthcare companies. This assistance can hopefully demystify stability practices for OTC monograph drug products and improve compliance during audits of the stability programs for these facilities.

Author Biographies

Mr. Jeffrey Needels has been working in the area of stability since 1992. He was first employed by Sandoz Pharmaceuticals in 1990 to work for the Laboratory Computer Operations group managing a LIMS system. In 1992, he moved to the research stability group in order to support computer applications and then shortly thereafter was appointed head of Research Stability in Lincoln Nebraska for Sandoz. In 2001, with the globalization of the now Novartis R&D group, he was appointed global head of Research Stability for the OTC business unit of Novartis Consumer Health. Jeffrey is functionally responsible for the global oversight of several development stability centers and for establishment of global stability practices. Jeffrey has a Masters Degree in Physical Chemistry from the University of Kansas in Lawrence.

Mrs. Mary W. Seibel is the Manager of Global Health Care Central Stability for Procter & Gamble in Cincinnati, Ohio. Her group is responsible for all R&D product commercialization for P&G's Oral Care and Personal Health businesses. This includes new product expiration dating studies, clinical studies and claims support studies for drug products, cosmetics, medical devices and dietary supplements. In addition, her group is responsible for the qualification of new raw material suppliers for ingredients used in our products. Her group is located in Cincinnati, Caracas, Venezuela, and Beijing, China. Mary joined Procter & Gamble in 1988 and has spent time working in various business units including Corporate Research, Pharmaceutical Discovery Analytical, Health Care BioAnalytical, and currently manages global Health Care stability programs. Prior to joining Procter & Gamble, Mary had experience in various clinical laboratories. She is a member of AAPS, CHPA, ASCP and participates in the Pharmaceutical Stability Discussion Group (PSDG). Mary received her B.S. degree in Biology from the University of Dayton in 1977.

Mrs. Karen Lucas is the Associate Director of Stability Management & Operations for Johnson & Johnson's Consumer and Personal Products Worldwide division in Morris Plains and Skillman, NJ. Her group is responsible for R&D new product and "post-market change" stability study management in support of product commercialization for Consumer cosmetic, OTC monograph drug products, formulated medical devices, NDA drug products and drug/device combinations. She also has responsibility for the stability chamber operations at the Morris Plains and Skillman NJ locations. She has been working with cosmetics, OTC monograph drug products, NDAs, Rx-to-OTC switches, and devices since she joined Pfizer Consumer Healthcare in 1986, and most recently joining Johnson & Johnson Consumer Worldwide in 2006. Karen has spent her time within Pfizer and J&J working in different business units including Product Development, Analytical/Technical Services, R&D/Marketed Product Stability, and currently as Associate Director of Stability Management & Operations. She is the chairperson of the CHPA Stability Working Group, steering committee and focus group member of the AAPS Stability Focus Group, member of the AAPS CMC Focus Group, ASQ and NJPQCA. She also participates in the Pharmaceutical Stability Discussion Group (PSDG). Karen received her B.S. Degree in Biochemistry from Cook College, Rutgers University.

Dr. Rachael Carlisle Roehrig joined CHPA in October 2007 and is the Director of Technical & Scientific Affairs for the Regulatory Scientific Affairs department. She is the coordinator for CHPA's Manufacturing Controls Seminar and the CHPA liaison to the Manufacturing Controls Committee and its three subcommittees; The Compendia Working Group, the Stability Working Group and the Nanotechnology Working Group. Rachael also participates on CHPA's behalf in Product Quality Research Institute (PQRI) activities in developing research projects to generate specific scientific information and in the International Conference on Harmonization (ICH) International Working Group on Quality. Prior to joining the CHPA team, Rachael had completed her graduate studies in the chemistry department at The Johns Hopkins University with a focus on inorganic photophysical chemistry, Rachael worked on several dissertation projects for her thesis entitled "From Light Harvesting to Intracellular Oxygen Sensing: Chromophores and Their Novel Applications in Photovoltaics and Molecular Cardiobiology." Along with Rachael's extensive graduate research experience, she joined the staff at Villa Julie College in Stevenson, MD as a physical chemistry lecturer and laboratory instructor focusing on both Thermodynamics and Quantum Chemistry.

Chapter 11

Challenges of Drug/Devices Pharmaceutical Products

Duu-Gong Wu

Abstract Due to significant differences in regulatory approaches and requirements by different review centers within FDA, design of a stability program to support the development and approval of a drug/device combination product can present significant challenges. Having an in-depth understanding of the stability issues and regulatory requirements associated with this type of drug/device combination products is critical to avoid the costly delay of the product development and marketing approval.

1. Introduction

In recent years, development of combination products, including drug/device and biologics/device combinations, is a new trend among pharmaceutical and device industries. However, meeting all requirements set forth by more than one review centers in US FDA, for example, Center for Drug Evaluation and Research (CDER), Center for Biologics Evaluation and Research (CBER) and Center for Devices and Radiological Health (CDRH) can represent many challenges. Often the manufacturers may not be familiar with the regulatory requirements for one part of combination product due to differences in approval process under different laws and regulations. Specifically, adequate stability testing of drug/device combinations to satisfy these requirements can be a difficult task. Without adequate stability testing, the integrity of drug/device combinations may not hold up during manufacturing, storage and transport or during use, resulting further delay of product approval. It is important to conduct stability studies properly to assure the stability of these combinations, starting from the early stages of drug testing and up to the post-marketing studies to assure its performance.

2. General Issues of Combination Products

2.1. Definition of Combination Products

As defined in 21 CFR § 3.2(e), the term combination product includes:

From: *Pharmaceutical Stability Testing to Support Global Markets*: Pharmasp,
Edited by: K. Huynh-Ba, DOI 10.1007/978-1-4419-0889-6_11,

1. A product comprised of two or more regulated components, i.e., drug/device, biologic/device, drug/biologic, or drug/device/biologic, that are physically, chemically, or otherwise combined or mixed and produced as a single entity.
2. Two or more separate products packaged together in a single package or as a unit and comprised of drug and device products, device and biological products, or biological and drug products.
3. A drug, device, or biological product that is packaged separately according to its investigational plan or proposed labeling is intended for use only with an individually approved specified drug, device, or biological product where both are required to achieve the intended use, indication, or effect and where upon approval of the proposed product, the labeling of the approved product ought to be changed, e.g., to reflect a change in the intended use, dosage form, strength, route of administration, or significant change in dose; or
4. Any investigational drug, device, or biological product that has been packaged separately according to its proposed labeling is for use only with another individually specified investigational drug, device, or biological product where both are required to achieve the intended use, indication, or effect.

Some examples for these different types of combination products are listed below.

- Physically or chemically combined products: drug-eluting coronary stents and pre-filled syringes filled with drug or biologic.
- Co-packaged products: needless injectors for use with drug or biologic.
- Separately packaged and cross-labeled products: a surgical kits and drug.

2.2. Regulatory Process

In the United States, the investigation and approval of devices, drugs and biologics for marketing are governed by different laws and regulations. For devices, there are Investigational Device Exemption (IDE) and PMA (Premarket Approval) processes for investigation and marketing approval of Class III devices and 510 K (Premarket Notification) for marketing of Class I or II devices. For the drugs or biologics, there are IND (Investigational New Drug Exemption) process for pre-clinical and phase I/II/III clinical investigations and NDA (New Drug Application) or BLA (Biologic License Application) process for marketing approval (see Table 11-1).

Due to the differences in process and data requirements by different review centers in FDA, the following regulatory issues need to be considered for the development of combination products:

- Product development life cycle.
- Review jurisdiction (RFD) and timeline.
- Clinical investigational phases.
- Submission format and data requirements.
- Labeling.
- Adverse event reporting.
- GMP and Quality System (21CFR210 211, 820).
- Manufacturing changes reporting.

Table 11-1. Regulatory process for devices, drugs and biologics.

	CDRH	CDER/CBER
Application for Clinical Trials	Investigational Device Exemption (IDE)	Investigational New Drug Exemption (IND)
Application for marketing approval	Pre-Market Approval (PMA, Class III Devices)	New Drug Application (NDA) or Biologic License Application (BLA)
Post-approval changes	PMA supplement	NDA or BLA Efficacy/ Manufacturing Supplement
Abbreviated pathways	510(k)	ANDA
	Pre-Market notification	Abbreviated NDA (N/A for BLA)

2.3. Drug and Device Development and Approval Process

To obtain device approvals, the first step is to determine the classification of devices under class I, II, or III and to develop the data (e.g., biocomparability, pre-clinical and clinical data). Depending on the classification of devices, different processes, ranging from Pre-Market Notification (also called PreMarket Clearance or 510 K) for Class I or most of Class II devices to IDE and Pre-Market Approval (PMA) for class III devices, are utilized. These approved devices are required to be properly labeled, registered and listed. Also, as a part of approval process, GMP inspections by FDA using quality system will take place at the manufacturing site(s). In general, the time required for product development, even with the class III devices, clinical investigation and approval of devices is shorter and it is a less expensive process relative to the development of drugs or biologics.

Development of drugs in general is a much longer and more expensive process involving many different stages from research and development, pre-clinical, and clinical studies. Human clinical studies are further divided into phase 1, phase 2, and 3 with different purposes and study durations and scales. These stages are designed to evaluate the product from initial safety and dosing determination to efficacy. The initial stage, starts with candidate selection, synthesis, and purification, followed by the animal testing, which would take up to approximately 18 months. The second stage of human clinical studies, starting from the phase 1 studies in healthy subjects to phase 2 and 3 in patient populations, takes somewhere between 2 and 5 years. Finally, submission and review of NDA/BLA for the marketing approval normally would require from 6 to 10 months.

Post-marketing surveillance, adverse event reporting and submission of post-approval changes are parts of the continuing regulatory oversights by FDA after marketing approval (see Fig. 11-1).

2.4. Review Jurisdiction for Combination Products

Currently, the jurisdiction with the lead review responsibility for combination products is assigned based on the product's primary mode of action (PMOA). For example, if the combination product contains a chemical drug component and a device and the drug is responsible for the PMOA, the overall product jurisdiction will belong to FDA Center for Drug Evaluation and Research

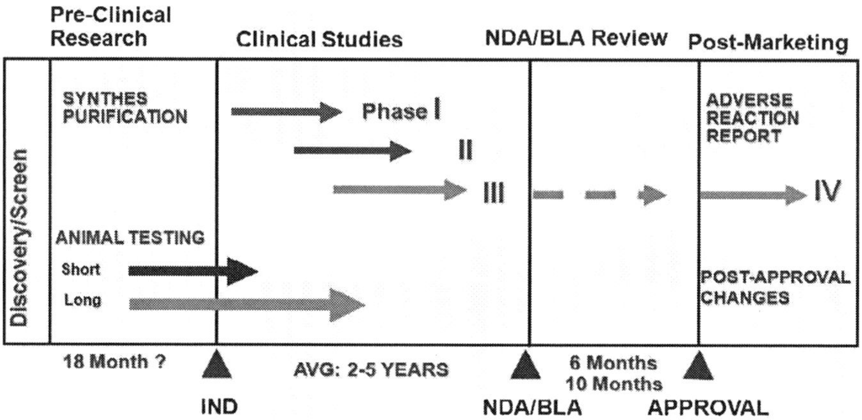

Fig. 11-1. Drug development and timeline.

Table 11-2. Lists of jurisdiction of products containing drugs in devices.

Product	Lead review center
Insulin/epinephrine/ injector pen-prefilled	CDER
Transdermal patches	CDER
Iontophoretic patches with approved drug	CDER and CDER
Wound dressing with antimicrobial agents	CDRH
Surgical kit w/ drug -with cross labeling	CDRH
Drug eluting coronary stents	CDRH

(CDER) in conjunction with the review of device component by the Center for Device. It must be discussed with the FDA early through a Request For Designation (RFD) process to determine which center should have the product jurisdiction and primary review responsibility. Such a determination would significantly affect the development program, including the time and costs required for the product approval. Table 11-2 lists some examples for the jurisdictions of various combination products.

2.5. General CMC Issues for Combination Products

For product approval, chemistry, manufacturing and controls (CMC) requirements for drug component in a combination product are similar to those for drug substance and drug product in a stand-alone drug product. However, additional data relevant to the unique characteristics of a combination product also need to be considered for regulatory approval. Typically, CMC information for a combination product, using drug-eluting stents as an example, should cover the following areas:

- Drug substance and formulation.
- Product characterization.
- Manufacturing process and in-process controls.
- Release specifications for products.
- Relationships of drug and mechanical function, durability or reliability of a device's components.

- Effects on drug stability by device components.
- Potential leaching of other components.
- Degradation products resulting from breakdown of the biomaterials from device components.
- Device impact on rate of release and absorption of drug.
- Drug stability in the combination products.

2.6. General Stability Issues

Stability is one of the most important characteristics of a pharmaceutical product. In addition to establishing expiry and assuring quality of a drug product, stability testing also serves many other purposes, including:

- To provide information on product characteristics.
- To understand its degradation pathways.
- To facilitate analytical methods development.
- To assist development of formulation and manufacturing process.
- To determine compatibility of container closure and delivery systems.
- To establish release and stability specifications.
- To determine in-use conditions.
- To link quality of pre-clinical and clinical batches to primary NDA stability batches for drug substance and drug product.
- To assess product comparability after manufacturing changes.

2.7. Legal Basis for Drug Stability Testing

Two main regulations under 21CFR provide the basis for the regulatory requirements of stability testing for drug products regulated by US FDA. GMP regulations 21CFR 211.166 specified that the manufacturers of drug products are required to (1) assess stability characteristics, (2) determine storage conditions and expiration dating, (3) provide establish written stability program including sample size, test intervals, storage conditions, specific methods, and container closure system, and (4) test stability under in-use and reconstitution conditions.

Furthermore, 21CFR 314.50(d) also indicates that information regarding stability specifications, analytical methods, stability data and proposed expiration dating are required for both drug substance and drug product in marketing applications.

These regulations mean that the following stability information is required to either comply with GMP regulations or to support the clinical investigations and marketing approvals:

- Stability results from drug batches used in pre-clinical and clinical studies during IND stages.
- Stability indicating tests and acceptance criteria.
- Primary and supporting stability data for marketing approval.
- Pre and post approval stability protocols and post approval stability commitment.
- Statistical analysis on stability data.
- Expiration dating and storage conditions.
- Stability evaluations for manufacturing changes.

3. Stability Issues for Combination Products

For drug component in a combination product, stability testing generally follows ICH guidelines (e.g., ICH Q1A). In comparison, requirements for stability testing on devices are very different. For example, aging studies of a device are not covered under current ICH guidance for drugs and biologics. Nevertheless, to assess the stability of a combination product, the tests should cover both drug and device components and the following issues need to be considered during development:

- Stability data requirements and shelf-life determination.
- Storage and shipping conditions.
- Device aging studies with the presence of drug components.
- Stability of carrier components for a drug, if applicable.
- Compatibility between drug and device components.
- Leachables and extractables during storage.
- Effects of manufacturing process on product stability.
- Sterilization process and product stability.
- In-use stability including mechanical stress.
- Stability testing and manufacturing changes.
- Reserve samples for stability testing.

There are also a few unique items which required special attentions during the design of a stability program for a drug-device combination, e.g., drug-eluting coronary stent. First, a shorter development cycle of a drug-device combination requires the final stability protocol to be developed early to collect sufficient data to support the final product approval. Second, unlike a traditional drug product, a combination product such as drug-eluting stent is typically manufactured at a small lot size, therefore matrixing and bracketing designs for stability testing are necessary to avoid consuming a large number of samples for the stability testing. Finally, stability indicating tests should include (1) in vivo and in vitro release rate for the drug component, (2) stability of critical inactive ingredients (such as integrity of coating polymers), (3) the interactions between the drug and the device, (4) effects of sterilization process on the product, and (5) monitoring of fatigue, corrosion, durability of the device part.

3.1. Stability Requirements for IDE and PMA

Based on the PMOA, FDA has assigned CDRH as the lead center for drug-eluting stent product (a combination product containing a Class III device); therefore, an IDE for human clinical studies and a PMA for marketing approval are required. For IDE submissions, the stability studies are designed to evaluate the stability of investigational formulation and device combination and collect the information for the development of final product; therefore, it is adequate to provide only preliminary data on representative materials to support the duration of clinical study.

However, for the approval of a PMA for a product such as drug-eluting coronary stent, it is necessary to collect stability data from the final product containing the final device design and formulation in the marketing package manufactured at the commercial production site and scale. The expiration dating of the products needs to be established based on the stability data.

Also, stability data collected under stress or in-use conditions may be needed for the final product approval. In a PMA submission for a product such as drug-eluting coronary stent, it is expected that the following information and data are provided in the application:

- Stability protocols including (1) description of packaging/container, (2) a list of analytical procedures, acceptance criteria and sampling procedures, (3) storage conditions(temperatures and humidity according to ICH guidance), (4) testing intervals and duration, (5) bracketing/matrixing protocols, if available, and (6) stability commitment (annual lots and action plan).
- Stability data and expiration dating for the commercial product.
- Data for one-time stability studies including (1) drug stability in coating solution, (2) in-use (e.g., after stent deployment) and shipping studies, (3) photosensitivity, and (4) product stability under sterilization and manufacturing processes.

3.2. Stability Indicating Methods

For a combination product such as drug-eluting coronary stent, in addition to the typical aging and functionality studies conducted for the device component, the following stability-indicating tests need to be performed to examine the stability of drug component:

- Assay (total drug load).
- Release profile (release kinetic).
- Impurities/degradation products.
- Moisture content.
- Polymer/carrier/matrix content.
- Sterility.

3.3. Sample Size and Stability Testing

Unlike a conventional dosage form such as oral tablet for which the batch size can range from few hundred thousand to multiple millions, the batch size for a combination product is in general very small (e.g., a few thousands for drug-eluting coronary stents). Stability testing for a combination product, if conducted under the same way for a conventional dosage form, will require the use of large number of samples, or possibly the entire batch. For example, the stability tests for a drug-eluting coronary stent product according to a typical protocol listed below in Table 11-3 will use more than 300 units for a single configuration (length and diameter). Much more units may be required if multiple configurations are manufactured. The stability program for a combination product, therefore, needs to incorporate matrixing and bracketing designs in order to reduce the sample size for stability testing.

Table 11-3. Stability protocol.

Temp\Time	**Month**						
	0	3	6	9	12	18	24
25°C/60% RH	X*	X	X	X	X*	X	X*
40°C/75% RH	X	X	X				

* Sterility Testing

Stability indicating tests at each timepoints listed on Table 11-3.

1. Assay/Purity/Impurities: 10 units.
2. Release rate: 10 units.
3. Moisture content: 10 units.
4. Sterility: 20 units (test at time zero, 12 months and 24 months.).

3.4. Timeline for Stability Studies

As previously discussed, a short timeline for the development of drug–device combination products requires the design of an expedited stability program. First, the information derived from previous stability studies conducted by the supplier(s) of the drug component should be fully utilized for the development of, for examples, a stable coating formulation for drug-eluting coronary stents, and stability-indicating analytical tests. Second, the finalization of stability protocol early enables the generation of sufficient data (time and lot numbers) to support a PMA approval. Third, an advanced planning for one-time studies such as force-degradation and in-use studies is also critical. Finally, it is important to avoid changes in the design of device component and the drug formulation (for example, the coating formulation for drug-eluting coronary stents) after pivotal studies. The requirements to repeat stability studies to confirm the product stability and expiry following such changes will unavoidably delay the product approval.

4. Conclusion

In the end, it is important to understand that the ICH stability requirements for drug have to be used in a combination product, especially for the manufacturers who are not familiar with regulatory requirements for drug components. Unlike the approval of devices for which expiration dating generally is not needed, an expiration dating supported by stability data is absolutely required for a drug-device combination product.

It is advised to design a stability program early enough to meet the short development timeline and to avoid frequent changes in the product and process. The stability indicating tests should be developed to cover unique characteristics of both drug and device components in a combination product and assess the effects on stability by the interactions between the two components. Also, it is a good idea to apply bracketing/matrixing and skip-lot approaches to reduce the sample size required for stability testing.

Finally, early interactions with regulatory agencies regarding RFD of the combination products and solicitation of their inputs during formal meetings or informal communications on the design of stability protocols are critical for a successful development program.

Acknowledgment

- Steve Koepke, Ph.D. SRK Consulting
- Stuart Portnoy, M.D. PharmaNet Inc.

Author Biography

Dr. Duu-Gong Wu is currently the Executive Director of Pharmanet, in its Regulatory Consulting Division. Prior to joining Pharmanet in 2004, He was with the US Food and Drug Administration for 12 plus years, worked as a reviewer, Chemistry Team Leader, and most recently the Deputy Division Director of Division of New Drug Chemistry II, Office of New Drug Chemistry (ONDC) in the Center for Drug Evaluation and research. His career at FDA encompassed a very unique combination of regulatory experience in both drugs of small molecules, peptides, botanical products and biotechnological/biological products. Dr. Wu served on many technical committees and working groups in FDA and CDER, including CDER Complex Drug Substance Coordinating Committee. He had represented CDER as a member of ICH Expert Working Groups for both Q5E and Common Technical Document-Quality for biotech products. Previously, he also participated in the drafts of ICH Q6B and Q5C for biotechnological drug products and many FDA CMC guidance. Dr. Wu completed his Ph.D. degree in University of Maryland School of Medicine and postdoctoral training in Johns Hopkins University School of Medicine.

Chapter 12

Practical Challenges of Stability Testing of Nutraceutical Formulations

Jairaj (Jai) Mehta

Abstract Nutraceuticals are a growing segment of the global consumer market. Until recently, it can be argued that the manufacturing and quality control of these products to include dietary supplements, botanicals, fish and animal based supplements, etc., has been largely unregulated in the United States. In addition to requirements from the compendia (USP, NF) which may not apply if the product is not labeled USP, the USFDA has recently (June, 2007) enacted regulations, which are not very clear on the requirements for expiration dating and supporting data and testing methods' validity. With this confusing regulatory requirements backdrop, the challenges for stability testing of nutraceutical formulations only get tougher, when one considers the complexity of multiple active ingredients, formulation and natural matrix effects and isolation of representative marker compounds. This article describes these challenges and attempts to provide a navigation compass to the pharmaceutical/nutraceutical scientist through this complex maze.

1. Introduction

Nutraceuticals have been quickly gaining attention due to the increasing consumer market share for wellness products. Thus, there has been a heightened focus in the marketing, claims substantiation, manufacturing, and FDA-based regulations of Nutraceuticals. A loose definition of Nutraceuticals would include herbs, minerals, vitamins, medical foods, dietary supplements, etc. The manufacturing of such products has been largely unregulated and historically has lacked desired quality control systems. It can be argued that up until now, the European countries have better regulated such products. The congressional attention to the widespread usage of nutraceuticals and the very real lack of quality standards practiced in the manufacturing industry for these products was finally reflected in the passage of the Dietary Supplement Health and Education Act (DSHEA – referred to herein after as the Act) in 1994. The United States FDA has recently, in 2007, issued a final rule for

From: *Pharmaceutical Stability Testing to Support Global Markets*: Pharmasp,
Edited by: K. Huynh-Ba, DOI 10.1007/978-1-4419-0889-6_12,

cGMP regulations of dietary supplements – a full 13 years after the passage of the Act. Due to the volume and breadth of the comments received in response to the proposed regulations, alas, the final regulations became very much diluted. The manufacturing landscape therefore is about to change. Along with these changes, better formulation and testing methods are getting attention. In contrast, the manufacturing practice recommendations in the USP for stability testing and the analytical methods validation are significantly different than the new FDA regulations. The industry scientists are just beginning to pay attention to the testing, methods and stability of Nutraceuticals as the focus on GMPs increase, and the playing field is already challenging with the USP versus FDA regulations. The task of stability testing of Nutraceuticals gets further challenging due to the potential interactions between the ingredients of various dosage forms, such as vitamins, herbal extracts and minerals in a complex matrix.

2. Regulatory Challenges

2.1. Dietary Supplement Health and Education Act (DSHEA): The Act

The Act was enacted by Congress on October 15, 1994. It amended the Federal Food, Drug and Cosmetic Act by adding section 402(g). Section 402(g) (2) of the Act provides, in part, the Secretary of Health and Human Services the authority to issue regulations for Good Manufacturing Practices (GMPs) for Dietary Supplements. It also requires that the standards may not be imposed for which there is no current and generally available analytical methodology. This caveat along with the requirement that the FDA pattern the GMPs for dietary supplements/nutraceuticals after the food regulations and not the drug regulations provided the FDA with limitations for enacting tough regulations.

2.2. Proposed Rule

In the slow and painstaking regulatory evolutionary process, it took the FDA almost 9 years after the congressional passage of the Act. to publish the proposed rule for Current Good Manufacturing Practice (CGMP) in manufacturing, packaging or holding dietary ingredients and dietary supplements on March 13, 2003.

There were no requirements for dissolution, disintegration, bioavailability, or expiration dating because scientific study was supposed to be still evolving. Especially for botanical dietary ingredients, few official methods were available to assess the strength of a dietary ingredient in a dietary supplement.

Nevertheless, the proposed regulations maintained that if the manufacturer were to use an expiration date on a product, data to support that date should be available. There should also be a written testing program designed to assess the stability characteristics of the dietary supplement, and use the results of the stability testing to determine the appropriate storage conditions and expiration dates.

Via the proposed regulations, the FDA also invited comment from the industry on whether any final dietary ingredient and dietary supplement CGMP rule should contain requirements regarding expiration dating and the feasibility of conducting tests to support such dates. The FDA also requested input regarding creating a different subset of products for which expiration dating would be required, for example, where the analytical method was fully characterized such as for vitamin, mineral and amino acid products only. Botanical or animal source dietary ingredients, where the analytical methods were still evolving could be exempted from the expiration dating requirements under this separation.

2.3. The Final Rule

On June 25, 2007, FDA Published Final Rule 21CFR Part 111 – Current Good Manufacturing Practice in Manufacturing, Packaging, Labeling, or Holding Operations for Dietary Supplements. The final rule applies only to the finished formulations of the nutraceuticals, and not to the dietary supplement ingredients that make up the dosage forms. Under this new cGMP final rule for the dietary supplement, the quality control departments of the formulation manufacturer are held accountable instead of the nutraceutical active ingredient manufacturer by this new (2007) or any other US FDA cGMP regulation to assure the quality of the dietary ingredients employed to manufacture the dosage forms. The final rule does get implemented over 3 years in a sliding scale based on the size of the manufacturing company, the smaller companies are benefited with more time to get acquainted to the new regulations.

21 CFR Part 111.70(e) of the new rule requires the establishment of finished product specifications for identity, purity, strength, and composition of the finished batch of the dietary supplement. For testing methods for these specifications 21 CFR Part 111.320(a) requires the methods to be appropriate, and 21 CFR Part 111.320(b) requires the identification and the use of appropriate, scientifically valid analytical methods for each established specification.

The FDA however, declined to define "test," "scientifically valid analytical method," or "scientifically valid method" in this final rule "because methods for components are evolving." According to the FDA, these definitions would become obsolete by the time they issue the regulations or by the time the methods are adopted in compendia.

Instead, the FDA maintains that it is the responsibility of quality control personnel to approve the use of those scientifically valid tests that will ensure a product's identity, purity, strength, and composition whether or not such tests are contained in a particular compendium.

Reference was made to an invitation for comments on expiration dating and feasibility of conducting stability tests needed to support such dates. According to the FDA, no persuasive comments were received to alter its position not to require establishment of expiration dating for dietary supplements. FDA also declined to offer guidance on the type of data that are acceptable to support an expiration date, other than to repeat themselves that any expiration date used, including a "best used by" date, should be supported by data.

3. Compendial USP Challenges

3.1. Legal Authority of USP over Nutraceuticals/Dietary Supplements

The United States Pharmacopoeia (USP) has been proactive in providing monographs on dietary supplements, as well as disintegration/dissolution methods and guidance on manufacturing practices for dietary supplements.

The Mission and preface Chapter of USP 30-NF25 (official August 1st, 2007–November 30th, 2007) provides legal basis for USP to address official standards for dietary supplements thusly.

The Dietary Supplement Health and Education Act of 1994 amendments to the FD&C Act name USP and NF as the official compendia for dietary supplements. The 1994 amendments also provide that a dietary supplement may be deemed misbranded if it is covered by a monograph in the USP or NF, is represented as conforming to this monograph, i.e., labeled as a USP product, but fails to conform, such that when tested, the product fails to meet the USP monograph standards.

The dietary supplement must therefore be represented (labeled) as conforming to a USP–NF dietary supplements monograph in order for the compendial standards to apply. This contrasts with pharmaceutical products, wherein conformance to the monograph is mandatory whether or not the product claims to conform.

3.2. Manufacturing Practices for Dietary Supplements: USP General Information Chapter <2750>

The General Information Chapter on Manufacturing Practices for Dietary Supplements states that many of the principles in this general information chapter are basic as that they apply equally to various types of products and levels of technology and that they are derived from the current good manufacturing practices for drugs. However, the practical application of these principles to dietary supplement may be different.

It is recommended by this chapter in the USP, for finished product specifications to include identity and strength testing prior to release. As for analytical methods, it is recommended to establish and document the accuracy, linearity, sensitivity, specificity and reproducibility when non-compendial methods are employed.

For stability testing, the USP recommends that there should be a protocol and the results should be to assess storage conditions and shelf-life assigned. The protocol should define storage conditions, test intervals and statistical acceptance criteria. Reliable, meaningful and specific test methods should be used, and accelerated studies with other supportive data could be used to predict shelf-life, provided stability studies are conducted concurrently under shelf-life conditions until the tentative shelf-life is verified.

As stated earlier, the FDA's Final Rule on CGMPs for Dietary Supplements was limited by the Act passed by the Congress in 1994, to be modeled after food GMPs, and also the FDA was hamstrung by the restriction imposed by the Act, not to promulgate regulation for nutraceutical formulations manufacturers to perform testing where the state of analytical science was not sufficiently advanced enough for practical everyday use. As a result of this dual mandate to the USP versus to the FDA by the Act, the confusion and dichotomy between the FDA regulations and the USP requirements, creates practical challenges for the manufacturer as to how to label the Nutraceutical product with expiration dating and test the stability of the product to support this expiration or "Use Before" date.

3.3. USP Verified Program for Dietary Supplements

USP's Verification Program is a multi-step process, fee for service program. First the company's quality system infrastructure is paper audited by USP, and then it is audited on-site against USP's & FDA's GMP requirements by USP auditors. After this audit, comes the step where random samples of the product, taken from the commercial stream (from the store shelves), are tested in the USP labs. Once the product is verified by the USP to meet with its monograph requirements, the "USP Verified" logo is available for marketing and labeling of the nutraceutical product. At this step, the product undergoes a market surveillance phase to ensure that it continues to retain ingredient strength and stability over its shelf-life.

In the author's opinion, if the manufacturers follow the extra quality standards of applying expiration dating on its Nutraceutical products based on the scientifically valid methods, validated according to the USP guidances, they should take advantage of the USP verification program for the added marketing differentiation from less conscientious manufacturers.

4. Analytical Methods/Technology Challenges

There are some products where test methods are readily available and they are scientifically sound, such as those for vitamins, minerals and vitamins with minerals. For some of these types of products, USP monographs exist and methodologies are available, and the methods used are relatively modern such as electrophoresis, HPLC, GC, etc. Unfortunately, some vitamins in the USP still have cumbersome inaccurate microbiology methods (e.g., those for cyanocobalamin, biotin, vitamin D, etc.).

There are even some cases of nutraceutical products, where method development and validation are challenging, such as identifying active ingredients in botanical products. Some of the high quality botanical nutraceutical manufacturers employ standardized extracts which are tested with well characterized marker components and well defined analytical methods such as USP, EP, WHO, AOAC, ANSI, AHP, etc. However, it is unknown in many instances if compounds other than the marker ones selected to quantitatively characterize a given botanical extract also are important for the pharmacological activity attributed to that botanical product. Thus, it is difficult to determine which ingredient must be routinely monitored. In addition, the ratios of active ingredients and relative potencies are also an issue not well understood, and it is still unknown how long those challenges will continue to exist, as well as how far, how quickly and how accurately the analytical science will go.

There are many herbal/natural products on the market, and more are constantly being added, where there are different combinations and sources (e.g., vegetarian/animal, geographic locations, etc.). For these types of varieties of products, it is difficult to develop assay methods even with the availability of marker compounds, let alone stability models or protocols. The challenges for the analytical scientists are therefore formidable, once the nutraceutical formulations manufacturer decides to step up to the "scientifically sound and valid" analytical methods requirement implied in the new FDA regulations to support an expiration or "Use Before" date, if placed on a label.

5. Formulation Related Challenges

The formulations of nutraceutical products can be very complicated, with many ingredients in the same capsule, tablet or liquid. A very good example of vitamins/minerals formulation challenges would be the change of an active ingredient source. Sometimes, the sources have different impurity profiles, leading to method interference with other vitamins. Another example would be the oxidative/reductive interaction of minerals with vitamins like Ascorbic Acid, Vitamin D, and Vitamin A, etc.

The use of coated vitamins and minerals to improve the shelf life by minimizing interactions could lead to poor analyte recovery. Trace quantities of vitamins or minerals in the presence of large quantities of other analytes or matrix would lead to poor recovery and detectability of the smaller quantity active ingredients.

As for shelf-life predictions, there are no clear-cut guidelines/science available, and Arrhenius relationships do not hold in most instances for nutraceutical products. The shelf-life prediction is challenging in some other cases also because of multiple active ingredients, each with its own stability sensitivities, and also the unpredictable effects of harsh accelerated conditions that the product is subjected to, which the product may never encounter in normal, commercial-lifetime on the shelf. Container-closure systems also have a large impact on the formulation stability, i.e., the unit dose blister vs. HDPE bottles, where the blister package, depending on the materials chosen, is significantly less protective in terms of moisture and oxygen permeation when compared to the HDPE container.

6. Conclusions

Compliance with regulatory requirements, even with the new FDA regulations (finalized in June 2007) for the dietary supplements does not require one to conduct a stability program or provide expiration/shelf-life dating. However, to establish a shelf-life, and apply an expiration or use-before date on a label, such a dating must be based on the supportive data arrived at using scientifically valid analytical methods.

As for USP requirements, the product should be compliant with USP monographs, if and only if the product carries the USP legend. The USP general information chapter on manufacturing practices of dietary supplements requires stability testing and shelf-life determination, and validation of the test methods used.

The analytical and formulation challenges have been attended to largely for some types of nutraceutical products such as vitamins. The challenges are still formidable for the analysts and formulation scientists in the manufacturing industry for the manufacture of herbal/natural/animal based nutraceutical products. The new FDA final rule and resultant regulations published in June 2007 make as best an effort as that allowed by the Congressional mandates and limits, to regulate nutraceutical/dietary supplement products, leaving ambiguities, and difficult choices for the manufacturer to comply with.

The USP verification program verifies conformance with USP and FDA draft regulations, and promotes customer confidence in your nutraceutical product formulation.

Author Biography

Mr. Jai Mehta has been in the pharmaceutical industry for more than 25 years. He has had a successful consulting business, and has held senior management positions at various pharmaceutical manufacturing firms. His responsibilities have included R&D and Regulatory functions, with oversight of Quality and Validation disciplines as well. Mr. Mehta had filed and received numerous Drug (ANDA, IND and NDA) approvals from the FDA during his career in the industry. Mr. Mehta is currently (2009) the Chair-Elect of the Nutraceutical Focus Group of the American Society of Pharmaceutical Scientists.

Mr. Mehta has audited and assisted bulk drug and dosage form facilities including Pharmaceutical and Nutraceutical products around the world for design of facilities and quality systems from either inception or current state of affairs, for compliance to FDA's cGMP standards. Mr. Mehta's expertise encompasses manufacturing facility design; product and process development and validation for bulk drugs and dosage forms which include sterile products including freeze-dried products, beta-lactam and oncology products, suspensions, solutions, and powders; topical semisolids and liquids; oral products like immediate release and modified release tablets, and capsules (hard and soft gelatin); and metered dose inhalers.

Mr. Mehta has a Bachelor's degree in Pharmacy and a Master's degree in Medicinal Chemistry from Bombay University and a Master's degree in Biopharmaceutics from the University of Illinois. He is a registered pharmacist in the states of Arizona and Illinois.

Chapter 13

Setting Tolerances for Instrument Qualification USP Chapter <1058>

Horacio N. Pappa and Kim Huynh-Ba

Abstract Qualification of analytical instruments has been a complex undertaking, not because the process of qualifying instruments is complex, but because several opinions abound on how to perform a successful qualification. Recently, the United States Pharmacopeia issued General Chapter <1058> to provide guidance for Analytical Instrument Qualification (AIQ). The draft chapter has been published in Pharmacopeial Forum (latest in PF 32:6) and has gone through multiple cycles of public comments. The chapter is scheduled to become official in 2008 (USP 31). The purpose of this chapter is to provide general guidance for the qualification of analytical instrument establishing a common terminology and defining roles and responsibilities of those associated with an instrument's qualification.

1. Introduction

Analytical instrumentation plays a critical role in the analytical information generation process. The USP has established several chapters to guide the process from the sample receipt to reportable data. These chapters include the use of weights and balances Chapter <41>, the use of volumetric apparatus Chapter <31>, validation of compendial procedures Chapter <1225>, verification of compendial procedures Chapter <1226>, and analytical data interpretation and treatment Chapter <1010> among others. These help the user assure the quality and the compliance of the analytical data generated.

The USP has developed the Analytical Instrument Qualification (AIQ) Chapter <1058>. This report summarizes the chapter and its recommended process for setting instrument qualification tolerances.

2. Analytical Instrument Qualification Process

Prior to the use of an analytical instrument for the generation of reportable data, it is imperative to provide documented evidence that the instrument is capable of performing its intended function. A qualified analytical instrument provides

From: *Pharmaceutical Stability Testing to Support Global Markets*: Pharmasp,
Edited by: K. Huynh-Ba, DOI 10.1007/978-1-4419-0889-6_13,

the foundation for the generation of quality analytical data. Analytical method validation ensures that the method used is capable of routinely meeting the required method acceptance criteria. System suitability tests and standards are controls to verify that at the time of analysis the method is performing in accordance with the acceptance criteria set forth in the procedure.

Instrument qualification is not a single continuous process, but instead the result of several discrete activities. For convenience, these activities are grouped into four phases: design qualification (DQ), installation qualification (IQ), operational qualification (OQ), and performance qualification (PQ).

2.1. Design Qualification

This qualification stage is performed prior to the purchase of the instrument. This step is a documented collection of activities that define the functional and operational specifications of the instrument, based on its intended purpose. For commercial off-the-shelf (COTS) instruments, it is typically done by the vendor and may encompass a broad range of applications. It is the responsibility of the purchaser to ensure that the instrument is suitable for the laboratory's intended applications, the manufacturer has in place a quality system for developing, manufacturing, and testing the instruments (e.g., power, humidity or vibration requirements), and the manufacturer and/or vendor adequately support installation, service, and training.

2.2. Installation Qualification

This stage occurs after the instrument is delivered to the user's site and must be performed prior to use of the instrument. and must be performed prior to use of the instrument. This step provides documented evidence that the equipment has been delivered as designed and installed appropriately. This qualification should include a system description, verification of instrument delivery in accordance with the purchase order specifications, verification of environmental conditions consistent with the vendor's requirements (e.g., power, humidity or vibration requirements), and network and data storage connectivity when needed. Documentation to verify proper assembly, installation, and functionality must be generated.

2.3. Operational Qualification

After a successful IQ, the instrument is ready for OQ testing. OQ is the documented collection of activities necessary to demonstrate that an instrument will function according to its operational specifications in its selected environment. Testing activities in the OQ phase may consist of measuring the instrument's non-changing parameters such as length, height, weight, or voltage inputs, testing secure data handling such as storage and backup and testing user-required instrument functions. Tables 13-1 and 13-2 shows an example of the qualification ranges expected for some key attributes of HPLC equipment and dissolution bath. OQ testing should be prospectively defined and successful completion documented.

2.4. Performance Qualification

PQ is also done at the user's site, and is intended to demonstrate that the instrument consistently performs according to the specifications defined by the user, and is appropriate for the intended use. In addition, the laboratory must

Table 13-1. Examples of expectations for HPLC.

Module	Attributes	General expectations
Pump	Flow rate accuracy	±2%
	Gradient accuracy	±1%
	Pressure tests	
Injector	Precision	1% RSD
	Linearity	$r \geq 0.999$
	Carryover	<1%
Detector	Wavelength accuracy	±2 nm
	Linearity	$r \geq 0.999$
	Noise and drift	10–5 AU; 10–4 AU/h
Column compartment	Temperature accuracy	±2°C

Table 13-2. Example of expectations for dissolution apparatus.

Parameter	USP tolerances	Generally accepted
Shaft wobble	No significant wobble	≤0.5 mm total run out
Basket wobble	±1 mm run out at the bottom of the basket	≤1.0 mm total run out
Shaft verticality	None	1.5° from perpendicular
Shaft centering	2 mm from center line	2 mm from center line
Speed	±4%	±1 RPM
Vibration	None	≤2 mil
Vessel temperature	±0.5°C	±0.5°C

conduct performance verification checks and perform preventive maintenance and repairs periodically to ensure that the instrument remains qualified. PQ testing and the appropriate preventive maintenance procedures and schedules should be prospectively defined and successful completion documented.

3. Roles and Responsibilities

Several different groups are typically responsible for the qualification of an instrument.

- The user is responsible for the instrument's qualification, operations, maintenance and on-going performance as well as assuring data quality, and appropriate training of personnel associated with the instrument's operation and maintenance.
- The quality unit reviews the AIQ process to assure it meets regulatory requirements. It is responsible for developing a sufficient understanding of the instrument qualification process and the instrument's function to properly execute this task.
- The manufacturers and developers are responsible for DQ, and should test the assembled instrument before shipping it to the user. Manufacturers and vendors should make available to the user a summary of their validation efforts, and notify all known users about the hardware or software defects

discovered after a product's release. The user is responsible for assuring that the manufacturer's DQ is consistent with the user's requirements.

4. Software Validation

Software used for analytical work can be classified into three categories: firmware; instrument control, data acquisition, and processing software; and stand-alone software.

Computerized analytical instruments contain integrated chips with low-level software (firmware). Firmware is therefore considered a component of the instrument itself. Thus, when the instrument is qualified at the user's site, the integrated firmware is also qualified. Data acquisition and processing software for many of today's computerized instruments is loaded on a computer connected to the instrument. Operation of the instrument is then controlled via the software, leaving fewer mechanical operating controls. The manufacturer should perform DQ, validate this software, and provide the users with a summary of validation. At the user site, holistic qualification, which involves the entire instrument and software system, is more efficient than modular validation of the software alone. Thus, the user qualifies the instrument control, data acquisition, and processing software by qualifying the instrument according to the AIQ process.

Validation of stand-alone software is not covered in Chapter <1058>. The validation process for this type of software is the responsibility of by the software developer, who also specifies the development model appropriate for the software.

5. Change Control

As with all cGMP's activities, changes that occur to analytical equipment need to be appropriately documented. Changes could occur as manufacturers add new features, improve current equipment or make changes to the instrument or its software, or as a result of instrument repair or maintenance. To implement any change, an assessment of the impact of the change of instrument performance must be performed and the instrument must pass the appropriate qualification stages to assure its continued acceptable performance.

6. Conclusion

It is critical to understand the importance of the analytical instrument qualification process and the steps it entails. A robust AIQ helps develop and maintain instruments that are reliable, perform as intended and are capable of generating analytical results that are consistent with user expectation. USP Chapter <1058> summarized the key steps in the process and provides general guidance to assist the user in assuring its successful completion.

Author Biographies

Dr. Horacio Pappa has been with USP since 2003. He is the scientific liaison to both the Expert Committee on General Chapters and the Expert Committee on Statistics. Horacio received his degree in Pharmacy from Buenos Aires University and earned a Ph.D. in Pharmaceutical Chemistry at the same university. Prior to joining USP, he worked in the pharmaceutical industry for 6 years in QA/QC. Horacio held the position of assistant professor of quality control in the Faculty of Pharmacy, Buenos Aires University, and Executive Secretary of the Argentine Pharmacopeia in the period 1997–2001. He is a Quality Engineer certified by the American Society for Quality.

Mrs. Kim Huynh-Ba is the Technical Director of Pharmalytik. She has over 24 years of experience in various analytical areas of pharmaceutical development and a primary focus in stability sciences, analytical development, outsourcing and technology transfer management. She has authored numerous technical publications and is frequently invited to present at national and international conferences. She is the editor of the "*Handbook of Stability Testing in Pharmaceutical Development: Regulations, Methodologies and Best Practices*", Springer.

Section III

Technical Concepts for Stability Program

Chapter 14

The Concept of Quality-by-Design

Mark A. Staples

Abstract In 2002, the FDA introduced a major initiative, recognizing that regulation of pharmaceutical manufacturing and product quality needed to be enhanced and modernized. In 2003, the International Conference of Harmonization (ICH) workshop participants proposed a new vision of pharmaceutical development – "The Desired State." The basic concepts of Quality by Design (QbD) as expressed in the ICH Q8 Guidance are described, as applied to pharmaceutical product development. QbD will ideally lead to better-understood products and manufacturing processes that will be subject to less variability in quality. ICH and FDA hope that the implementation of QbD will increase regulatory flexibility, creating an easier path for the manufacturers to introduce process and product improvements.

Perspectives are provided on how QbD may affect design, implementation and reporting of pharmaceutical stability testing in the global marketplace.

1. Quality by Design: Introduction

In 2002, the FDA introduced a major initiative, recognizing that regulation of pharmaceutical manufacturing and product quality needed to be enhanced and modernized (FDA 2004a). At the ICH meeting that took place in Brussels, Belgium in July 2003, (ICH 2003), an agreement was reached on a common vision and approach to achieve a harmonized pharmaceutical quality system that would be incorporated into the Pharmaceutical Development Section (3.2.P.2) of the regulatory submission format for drug products as described in the ICH M4 Common Technical Document (CTD) guidance document (ICH 2000). This vision is applied throughout the product lifecycle, and it emphasizes an integrated approach to risk management, and sets the goal, "The Desired State" (Hussein 2005), as the situation when product quality and performance are achieved and assured by the design of effective and efficient manufacturing processes (FDA 2004b).

Subsequently, ICH Guidance Q8 (Pharmaceutical Development) was approved (ICH 2005a), and related Guidances involving Risk (ICH Q9) (ICH

From: *Pharmaceutical Stability Testing to Support Global Markets*: Pharmasp,
Edited by: K. Huynh-Ba, DOI 10.1007/978-1-4419-0889-6_14,

2005b) and Quality Systems (ICH Q10) (ICH 2007a) were created. The Annex to ICH Q8 (ICH 2007b) more fully describes several concepts, including Design Space and statistical tools such as Design of Experimentation (DOE).

These ICH Guidances focus on product and process knowledge, and are particularly important in terms of increased regulatory flexibility with regard to product characterization, process characterization, and also with regard to the specific requirements of a particular dosage form (FDA 2004c). However, ICH Guidances Q8, Q9 and Q10 increase the stringency of requirements for manufacturers to clearly and concisely communicate the rationale behind the product and process design, and to justify subsequent changes during the product lifecycle. Application of QbD may affect choices of processes, formulations, and specifications, changing the inputs to stability programs. However, general practice and stability study guidances are not directly affected by QbD. Risk as a QbD-related consideration may stimulate the broadening of stability programs by the deliberate and strategic addition of more formulations and dosage forms to provide back-ups to lead choices, or perhaps a wider range of environmental test conditions to better specify the design space.

2. Quality by Design: Implementation

The ideal system attributes of a QbD-based development process (Nasr 2006) are in principal straightforward and make practical sense. Figure 14-1 presents a schematic of QbD stages (visualizing the initial product design as the tip of the pyramid and process monitoring as the base). The QbD stages include: (1) design the product to meet patient requirements (“unmet medical needs”), (2) design the process to consistently meet product critical quality attributes, (3) use the quality system to integrate product and process knowledge gained during development, (4) understand the impact of formulation components and process parameters on product quality, (5) identify and control the criti-

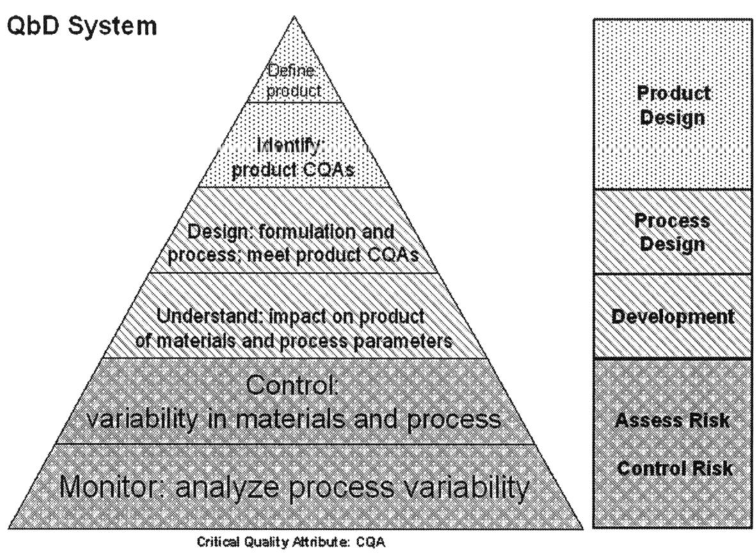

Fig. 14-1. QbD system: stages

cal sources of process variability, and (6) continually monitor and update the process to assure consistent quality over time.

QbD differs from the typical development process for pharmaceutical (or biotechnological) products, partially because it involves leveraging prior product knowledge to support more deliberate design effort from product conception through commercialization. Systematic use of tools (i.e., DOE: a structured, organized method for determining the relationship between factors affecting a process and the response of that process (Chen and Moore 2006)) reduces empiricism in creating the project knowledge base (Poochikian 2005). In particular, reliance on performance of test batches and historical data to establish specifications is de-emphasized in favor of identifying critical process parameters and control limits as early as possible in development, and then designing the product to meet those critical quality attributes. Greater predictability of the impact of process variables on product quality should lead to better product quality assurance. Dr. J. Woodcock of the FDA sees this process culminating in clinical studies that would then be viewed as confirmatory performance testing of the model (Woodcock 2005a). Figure 14-2 (FDA 2007) depicts the FDA model of QbD, encompassing continuous improvement via adjustments to product and process throughout the product lifecycle.

QbD will likely be more challenging to apply to biologically derived products than to chemically synthesized drugs, due to the greater complexity of these products (De Palma 2006; Kozlowski 2007). The FDA seems to be supporting this assumption by more gradually phasing-in QbD principles to the biotechnology industry (Winkle 2007).

The project database, including all product and process data, allows specification of a design space that includes all process parameter changes that may be made while maintaining acceptable product quality (Rathore 2007). Figure 14-3 illustrates the various layers of a Design Space, and provides several examples

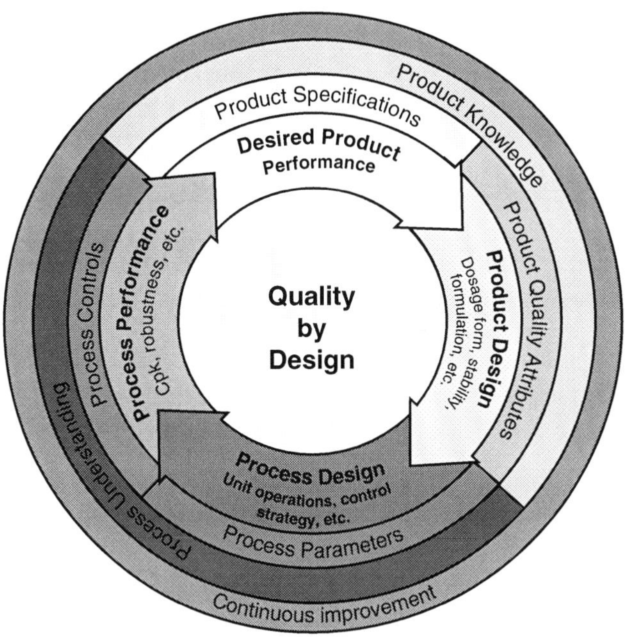

Fig. 14-2. FDA model for quality by design

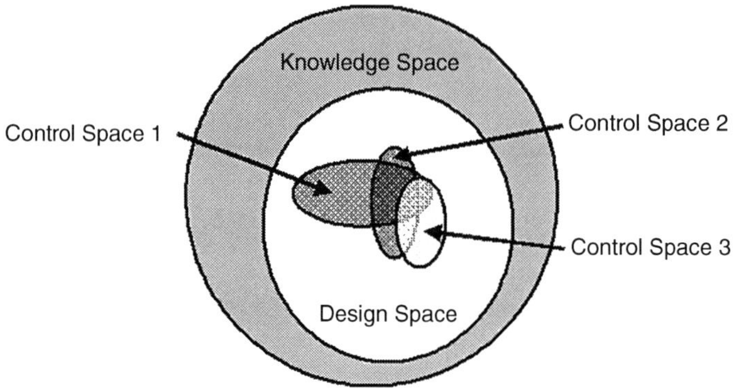

Fig. 14-3. Design space illustration

of Control Spaces that represent different combinations of parameters that still result in acceptable quality product. Knowledge space extends beyond the design space and includes conditions that lead to unacceptable product quality (for example, from stress testing).

3. Quality by Design: Benefits of The Desired State

The Desired State describes a situation where everybody (patients, manufacturers, and regulatory officials) wins: "It is a maximally efficient, agile, flexible pharmaceutical manufacturing sector that reliably produces high quality drug products without extensive regulatory oversight" (Woodcock 2005b).

Regulators win because they will review submissions which are more readable, briefer, and understandable, that highlight the science of the product. The burden on regulators should also decrease if there are fewer corrective actions necessary during the product's lifecycle as a consequence of correctly applying QbD and developing a quality product. Manufacturers will have an easier path to improve their product and should experience lower costs due to fewer compliance actions, penalties and recalls. Patients will receive a higher quality product that may also be commercially available sooner (Power 2007).

The new regulatory approach outlined in ICH Guidances Q8, Q9 and Q10 should produce several benefits (Nasr 2007): (1) faster review cycles, (2) greater likelihood of assembling an approvable initial application, (3) replacement of some release testing with process monitoring, and (4) risk-based regulatory decisions that facilitate continuous improvement of process and product.

4. Quality by Design: Implications for Stability Testing

QbD will have an impact on the design, implementation and reporting of stability testing, although the primary impact will be on the product design, manufacturing and testing that precedes stability studies.

Stability tests are stability-indicating measures of critical product quality attributes over time. The product must be designed (formulation, packaging, etc.) such that it will continue to meet critical quality attributes for a period of time that meets the patient needs. Similarly, the process must be designed (parameters, controls) to yield product that retains acceptable quality throughout the product shelf life.

The QbD concept naturally accommodates stability considerations. Although the design space definition may initially focus on process variable impact on product quality at time of release, maintaining acceptable quality over the target shelf life of the product must be built into requirements for an acceptable design space. Accordingly, the design space must be defined as the overall universe of possibilities within which, process parameters may be varied without compromising product quality *over the shelf life of the product.* This requirement implies that stability study design and implementation affects QbD as well as the reverse.

Adding acceptable stability as a requirement should be expected to affect the design space. For example, certain process parameters may need to have a narrower range of values in order to provide adequate assurance that product will not go out of specification prior to the expiration date, although product quality at time of release may be satisfactory. As is true without additional QbD-related constraints, the formulation and/or dosage form may need to be adjusted in order to meet the target stability requirements and post-release testing.

ICH Q9 (ICH 2005b) acknowledges risk as an element of program design. Risk-based stability programs should incorporate an assessment of a variety of back-up product configurations in the event of long term, unanticipated, adverse effects detrimental for the occurrence of product quality. Many variants of product composition, of product manufactured by different processes, and of product packaged in different container-closure systems can be maintained in controlled storage conditions. Since testing need not be done at frequent intervals, a large number of samples do not imply correspondingly large testing costs and labor. Also, since additional samples may be drawn from research experiments, it is not mandatory to have a high production cost to make the samples. When included in the stability program, these alternative product configurations can provide real-time stability data of demand. When a problem surfaces with the lead prototype product, knowledge of various redesign options can save significant project development time and money.

QbD may benefit stability programs, because an increased understanding of product stability and enhanced product quality should enable the initial submission of less extensive stability data. Stability properties may be explained by principles derived from the database as opposed to presenting reams of tabulated historical data to support shelf life.

References

Chen C-W and Moore C (2006) FDA/Industry Statistics Workshop, Washington DC, Role of statistics in pharmaceutical development using quality-by-design approach – an FDA perspective

De Palma A (2006) PharmaManufacturing.com, QbD takes Biopharma by storm

FDA (2004a) Pharmaceutical CGMPs for the 21st Century – a risk-based approach (SEP04)

FDA (2004b) Pharmaceutical CGMPs for the 21st Century – a risk-based approach; ICH Q8 EWG reference to The Desired State, p. 13 (SEP04)

FDA (2004c) ONDC's new risk-based pharmaceutical quality assessment system (29 Sep 2004)

FDA (2007) Pharmaceutical CGMPs for the 21st Century – a risk-based approach: Progress Report, Appendix 19 (May 07)

ICH Harmonized Tripartite Guideline (2000) M4Q: The CTD – Quality (NOV2000)
ICH Harmonized Tripartite Guideline (2005a) Pharmaceutical development Q8 (Nov 2005)
ICH Harmonized Tripartite Guideline (2005b) Quality risk management Q9 (Nov 2005)
ICH Draft Consensus Guideline (2007a) Pharmaceutical Development. Annex to Q8 (Nov 2007)
ICH Harmonized Tripartite Guideline (2007b) Pharmaceutical Quality System Q10 (May 2007)
Hussein A (2005) PharmaManufacturing.com, The Desired State: PAT and the Road to Enlightenment
Kozlowski S (01 Oct 2007) Protein therapeutics and the regulation of quality: a brief history. Int Biopharm
Nasr MM (2006) FDA/Industry Conference, risk-based CMC review and quality assessment: what is Quality by Design (QbD)?
Nasr MM (2007) FDA quality initiatives workshop, Quality by Design (QbD) – a modern system approach to pharmaceutical development and manufacturing – FDA perspective
Poochikian G (2005) PQRI Workshop: Leachables and Extractables, Best Practices Recommendations: Regulatory Science Strategies
Power M (2007) Insight (BearingPoint) quality by design: life sciences companies need to step up now
ICH Steering Committee Press Release, Brussels (2003) Step 4 for the ICH6 Program (JUL03)
Winkle HN (2007) PDA/FDA Joint Regulatory Conference, Implementing Quality by Design; related press release (26OCT2007) in The MQN Weekly Bulletin, Biotech QbD Pilot Program to Start Within Two Months
Woodcock J (2005a) The concept of pharmaceutical quality. Industrial Pharmacy, pp. 2–9, (Mar 2005)
Woodcock J (2005b) AAPS Workshop –Pharmaceutical quality assessment – a science and risk-based CMC approach in the 21st century, Pharmaceutical quality in the 21st century – an integrated systems Approach
Rathore AS, Branning R, Cecchini, D (01APR2007) BioPharm Int, Quality: Design Space for Biotech Products

Author Biography

Dr. Mark Staples received his BA in 1975 and his Ph.D. in 1979, both in Biochemistry from the University of Kansas. He was a Postdoctoral Fellow at Harvard Medical School 1979–1980 and held a part-time Research Appointment in 1982–1986. His research emphasized structure-function relationships of peptides and proteins. For more than 20 years Dr. Staples has helped to scientifically transform interesting projects into commercial products, serving unmet medical needs. His product development activities include pharmaceutical development of proteins and peptides and project management of the technical (CMC) portion of regulatory submissions. He has participated in five early stage companies in addition to founding and leading the Biogen Formulation and Analytical Development Group from 1988 to 1997. His titles were at the Director and Vice President levels for ten years from1997 to 2007. In 2007 he founded Cusp PharmaTech Consulting LLC in Cambridge, MA.

Dr. Staples was the Chairperson of the BIOTEC Section (2006). He has also chaired the AAPS Newsmagazine Content Advisory Board and Subcommittee and was Membership Chair of the AAPS BIOTEC Section 2001–2003. He has served on the Board of the Parenteral Drug Association New England Chapter (2005–2006), was President of the Chapter in 2003–2004 and Treasurer 1993–2002, and in March 2007 he received the PDA Chapter Volunteer Award.

Chapter 15

Forced Degradation and Its Relation to Real Time Drug Product Stability

Steven W. Baertschi

Abstract Forced degradation (stress testing) studies are fundamental to the developing stability-indicating analytical methods as part of the overall analytical control strategy. Stress testing studies conducted on both the drug substance and product are useful for gaining an understanding of the degradation chemistry of a drug. The degradation products observed in such studies are "potential" degradation products that may or may not be formed under relevant storage conditions, but these products reveal the degradation pathways available to the drug and facilitate the development of stability-indicating analytical methods. These methods can then used to monitor drug stability during long-term stability studies to determine which products actually form under room temperature and accelerated stability conditions. In this presentation, the foundational aspects of forced degradation studies and their relationship Quality by Design principles will be presented along with critical aspects predicting the "real time" drug stability and degradation profiles will be discussed.

1. Introduction

"Stress testing of the drug substance can help identify the likely degradation products, which can in turn help establish the degradation pathways and the intrinsic stability of the molecule and validate the stability indicating power of the analytical procedures used." This is the definition provided by the International Conference on Harmonization (ICH) guidance on stability testing (International Conference on Harmonization 2003). This is an important definition because it distinguishes stress testing (also known as "forced degradation") from two other types of stability-related studies: accelerated stability testing and long-term stability studies. Thus, stress testing is a predictive research tool that is used to *discover* stability issues with a drug molecule, providing the scientific foundation for developing stability-indicating methods. Accelerated stability testing and long-term stability testing can then utilize the stability-indicating methods in a more formal manner to generate the specific

From: *Pharmaceutical Stability Testing to Support Global Markets*: Pharmasp,
Edited by: K. Huynh-Ba, DOI 10.1007/978-1-4419-0889-6_15,

and quantitative information needed for estimation/determination of shelf-life as part of a regulatory submission.

2. Background and Definitions

2.1. Foundation of Stability Testing

The ICH guideline for stability testing of new drug substances and products (Q1A(R2)) states that the stress testing should take place to determine the "intrinsic stability" of the drug molecule. While the specific meaning of "intrinsic stability" is not delineated in the guidance, the concept can be thought of as encompassing four main aspects: (1) conditions leading to degradation; (2) rates of degradation (relative or otherwise); (3) structures of the major degradation products; and (4) pathways of degradation. The following "stress" conditions are mentioned in the guidance: elevated humidity, elevated temperature, photostability/photolysis, oxidation, and susceptibility to hydrolysis across a wide pH range.

2.1.1. Relevant Conditions for Stress Testing Studies

The design of stress testing studies should take into consideration these aspects: manufacturing conditions, analytical handling, storage and distribution, and patient "in use". Since conditions can vary substantially depending upon the location of storage facilities and means for distribution/transport, it is usually best to anticipate the worst-case realistic conditions that might be experienced in storage and shipping. Consideration should also be given to how patients might reasonably be expected to handle (or mishandle) their medicine once it is in their hands. Thus, the design of stress conditions should anticipate temperature excursions during shipping (Zahn 2006), elevated humidities (Zahn et al. 2006), exposure to direct or window-filtered sunlight (and not simply cool white fluorescent light), and exposure to oxidation sources such as peroxides (e.g., from excipients) and other autoxidation initiating sources (e.g., radicals, transition metals).

3. Design of Studies

3.1. Conditions for Stress Testing to Predict Degradation

Stress testing needs to be conducted to investigate four main degradation pathways of a typical drug. These pathways are:

- *Hydrolytic:* this can be accomplished by storing the drug in solutions or suspensions from pH 1 to 13, up to a temperature of 70°C for up to 1 week.
- *Thermolytic:* this can be accomplished by storing the drug at temperatures up to 70°C in both high humidity conditions (75%) and low humidity conditions (e.g., 10–20%), typically for 3–6 weeks.
- *Photolytic:* this can be accomplished by exposing the solid and solutions to both visible and UV light in excess of ICH minimum confirmatory exposure (twofold to fivefold excess recommended) (International Conference on Harmonization 1997). This can be done by either of two options: (1) exposing the drugs to a combination UVA lamp and cool white fluorescent (Option 2), either sequen-

tially or simultaneously, and (2) exposing to simulated solar radiation (Option 1 source, e.g., xenon or metal halide) behind window glass.[1]

- *Oxidative:* oxidative degradation can be difficult to mimic because of the different reactive oxygen species and complex oxidative mechanisms that are possible.

3.2. Predictive Oxidative Susceptibility Testing

There are three major oxidative pathways in drug degradation; autoxidation, peroxide-mediated, and electron-transfer mediated. Stress testing studies designed to mimic these pathways are presented here as "Predictive" oxidative susceptibility tests. Oxidative pathways from exposure to other reactive oxygen species (e.g., ozone, hydroxyl radicals, singlet oxygen) are possible, but less common. Such possible tests will be briefly discussed below as "Investigative" oxidative susceptibility tests.

3.2.1. Autoxidation

Autoxidation is a radical-initiated process where the source of the initiation is often not well understood. Initiation could be from thermal homolysis of organic peroxides, from heavy metal ions, from photochemical reactions, or even ozonolysis. Once a radical is formed in the system, reaction with ground state molecular oxygen is fast, leading to hydroperoxy radicals and subsequently to propagation of the chain reaction (Fig. 15-1).

Since the rate-limiting step in autoxidation is radical formation, the conditions needed to mimic autoxidation involve the exposure of the drug molecule to radicals formed in a controlled manner in the presence of molecular oxygen. Thus, the AIBN (azobisisobutyronitrile) test designed by Boccardi (Boccardi et al. 1992) is now widely used and has proven to be highly effective at initiating autoxidation processes that mimic long-term oxidative processes.

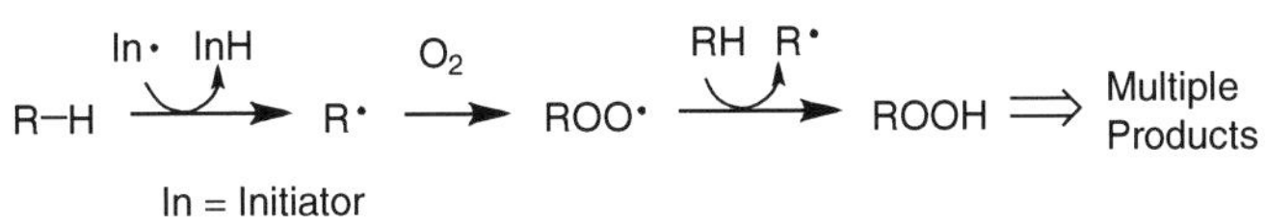

$RH + In^{\bullet} \rightarrow R^{\bullet} + InH$ **Initiation**

$R^{\bullet} + O_2 \rightarrow ROO^{\bullet}$

$ROO^{\bullet} + RH \rightarrow ROOH + R^{\bullet}$ **Propagation**

$2\ ROO^{\bullet} \rightarrow ROOOOR$

$ROOOOR \rightarrow$ non-radical products **Termination**

Fig. 15-1. Basic steps involved in the autoxidation process

[1] For reference, the total UV radiation from the sun on a clear day in Miami, Florida can be as high as 475 Wh/m^2 (integrated UV radiation from 295 to 400 nm). Data is from the Eppley total UV radiometer, July 24, 1996, courtesy of Norma Searle, consultant, and Atlas Electric Devices, Chicago, IL, personal communication.

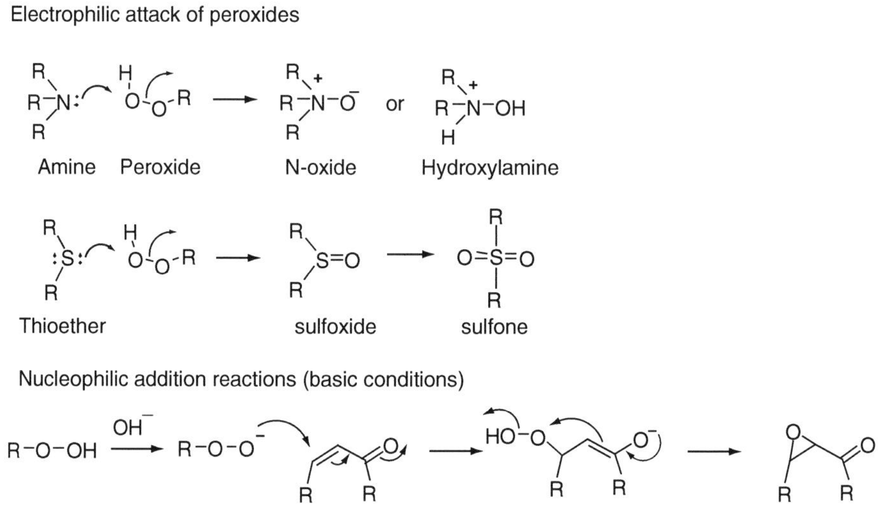

Fig. 15-2. Electrophilic (*upper*) and nucleophilic (*lower*) reactions of peroxides illustrated with amines, thioethers, and α,β-unsaturated systems

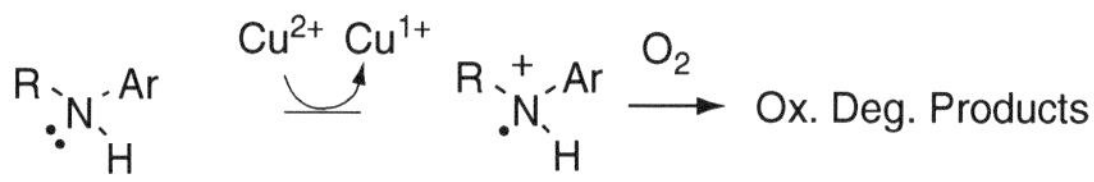

Fig. 15-3. Scheme illustrating a transition metal (copper II) catalyzed oxidation of an aromatic amine-containing drug

3.2.2. *Peroxide-Mediated*

The second major pathway is peroxide-mediated oxidation. Peroxides can be found at varying levels in certain excipients (e.g., povidone, polyethylene glycol, polysorbates, and hydroxypropyl cellulose) (Wasylaschuk et al. 2007) and the levels may increase over time due to autoxidation processes. Deliberate exposure of drug substances to peroxides can be accomplished simply by dissolving the drug in dilute aqueous hydrogen peroxide (e.g., 0.3–3%). While it is common for pharmaceutical companies to do this exposure at elevated temperatures in order to "speed up" the oxidative processes, this is not recommended due to the increasing potential for formation of highly reactive hydroxyl radicals that will aggressively oxidize many compounds in unrealistic pathways. Instead, it is recommended to do the peroxide exposure at <40°C for 1–3 days. Such conditions allow the non-radical peroxide oxidative mechanisms to dominate (e.g., electrophilic attack and nucleophilic additions, see Fig. 15-2), permitting more realistic predictive assessments of degradation caused by peroxides.

3.2.3. *Electron Transfer*

Electron transfer mechanisms (e.g., those involving the removal of an electron from a functional group within a drug molecule) can occur as a result of binding to and reaction with transition metals such as copper (II) and iron (III). In the process, the oxidation state of the metal is reduced and the drug molecule can become an unstable radical cation, readily reacting with molecular oxygen to form oxidative degradation products as illustrated in the case of copper (II) for an aromatic amine-containing compound in Fig. 15-3. Electron transfer mechanisms can also occur via direct reaction with molecular oxygen if there are functional groups in the molecule that are very electron-rich (e.g., phenolate anions) (Nishinaga 1980).

3.3. Investigative Oxidative Susceptibility Testing

As mentioned above, oxidative degradation can be very complex, and it can be useful to have in one's arsenal other oxidative reagents/techniques to allow for rapid investigation of oxidative degradation problems. For example, the use of peroxides that are stronger than hydrogen peroxide (e.g., peracetic acid, meta-chloroperoxybenzoic acid, or Oxone™) can often yield significant levels of oxidative products in less than 30 minutes. Another useful tool is exposure of the drug to singlet oxygen. A simple system to produce singlet oxygen involves the use of photosensitizers such as rose bengal or methylene blue in the presence of visible light, again typically for less than 30 minutes (Boccardi 2005). Singlet oxygen will often produce some of the same products formed by other oxidative mechanisms, and thus can be useful for producing larger quantities of select degradation products for identification purposes. Further, singlet oxygen-derived degradation, while unusual, is not unprecedented (Brustugun et al. 2000; Seburg et al. 2006; Reed et al. 2003).

Other oxidative reagents such as Fenton's reagent (Fe^{2+} in the presence of hydrogen peroxide) to form hydroxyl radical, a very aggressive oxidant, can in some cases be useful to rapidly produce certain oxidative degradation products in a manner of seconds (Baertschi et al. 2006).

3.4. Drug Product Stress Testing

Stress testing studies should also be conducted on the formulated product because drugs can and often do degrade differently in specific dosage forms and in the presence of specific excipients. For solid oral dosage forms (e.g., capsules and tablets), stress testing can involve exposure to elevated temperatures, humidities, and to light (i.e., photostability testing (Thatcher et al. 2001a, b)). For lyophilized powder products, the same conditions are recommended plus reconstitution studies at elevated temperature and in the presence of light. For liquid formulations, in addition to exposure to elevated temperature and humidity and to light, acid/base hydrolysis studies are recommended because of the potential for changes in the degradation profile as a function of the pH and the components of the liquid formulation.

4. Using Quality-by-Design Concepts

4.1. Quality by Design (QbD) Principles

QbD is defined by ICH as a "systematic approach to pharmaceutical development that begins with predefined objectives and emphasizes product and process understanding based on sound science and quality risk management" (International Conference on Harmonization 2006a, b). The concepts of QbD can provide a useful construct for understanding how stress testing can provide the scientific foundation for analytical method development and the development of control strategies for stability. Thus, a drug degradation (or stability) "design space" can be envisioned in the context of knowledge gained through scientifically sound and thorough stress testing.

The base of knowledge that can be acquired through stress testing can be thought of as a "knowledge space", which can be related back to the concept of "intrinsic stability" used by ICH in the definition of stress testing (see above). The integrity of the degradation knowledge space is dependent on the

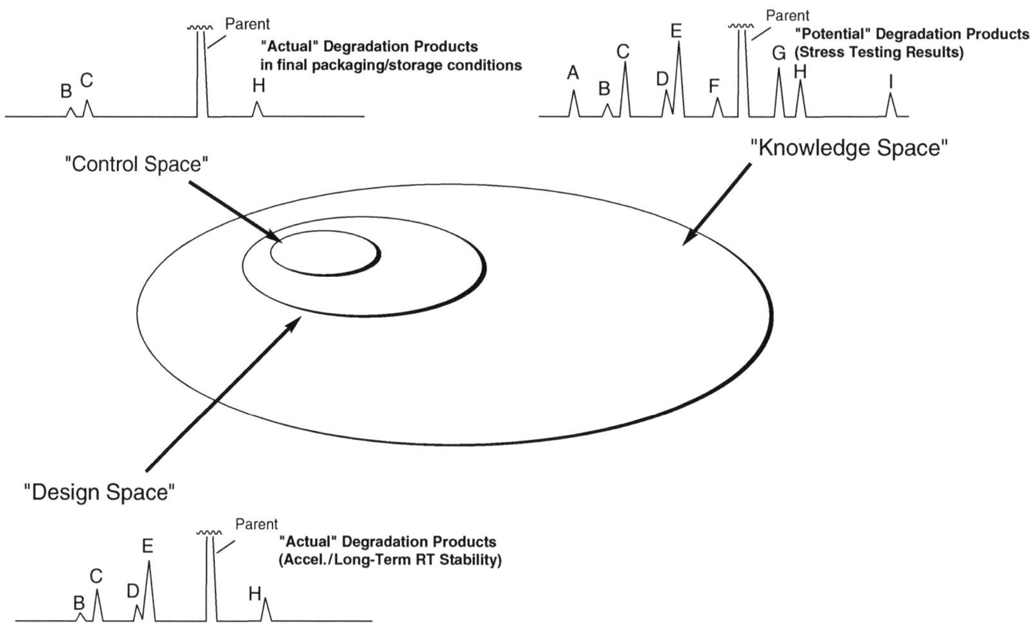

Fig. 15-4. Conceptual illustration of the application of QbD-type concepts to stress testing and its place in the development of stability-indicating analytical methods and control strategies

scientific quality of the investigation of all likely modes of degradation from stress testing under relevant conditions. More simply put, stress testing should identify all the reasonably possible degradation products that can result from “real-world” conditions; such products can be thought of as “potential” degradation products. The “actual” degradation products that are formed for a particular drug will be dependent on the physical state, the dosage form, packaging, and storage conditions, and should be a subset of the potential degradation products. This set of products can be thought of in terms of “design space” in a QbD construct, and the construct can be further refined to have a “control space” that is a subset of the design space. This construct is illustrated in Fig. 15-4.

4.2. “Holes” in the Degradation Knowledge

As illustrated in Fig. 15-4, the integrity of the design and control spaces is dependent upon a complete degradation knowledge space from stress testing. If there are gaps or missing pieces of information in the knowledge space, these gaps can be conceptualized as “holes” in the cartoon representation of the knowledge space. It is important to understand what some of these holes are so that they can be avoided.

Incomplete or poorly designed or performed stress testing studies can result in the lack of detection of a significant degradation pathway, and therefore an incomplete understanding of the degradation chemistry of the drug. Full coverage of hydrolytic, oxidative, thermolytic, and photolytic stress conditions is critical to make sure that the degradation knowledge space is complete. An important part of this “coverage” is the analytical methodology that is used to separate and detect degradation of the parent drug and the resulting products (Olsen and Baertschi

2003; Baertschi 2006; Jansen et al. 2005). Poor analytical methodology can result in lack of separation of degradation products from the parent, from each other, or lack of detection altogether (e.g., from poor choice of detector, non-elution, volatility, or poor chromatography) (Baertschi et al. 2005).

It is also critical to remember that the formulation and packaging materials can result in degradation pathways not present in the drug substance alone. This can result from a variety of sources including drug-excipient reactions, impurities from excipients or packaging, (Waterman and Adami 2003) changes in protonation state induced by the formulation, (Zannou et al. 2007) formation of amorphous content during the formulation process, or deliquescence lowering caused by the formulation (Guerrieri et al. 2007; Salameh and Taylor 2005).

4.3. "Degradation Design Space" Changes

If there are holes in the degradation knowledge space, there is the possibility that the design space might also contain holes, or that any changes in the design space location on the knowledge space surface might result in the discovery of holes in the design space, i.e., new, unanticipated stability problems or issues (see Fig. 15-5). As shown in Fig. 15-5, a change in the formulation of a drug

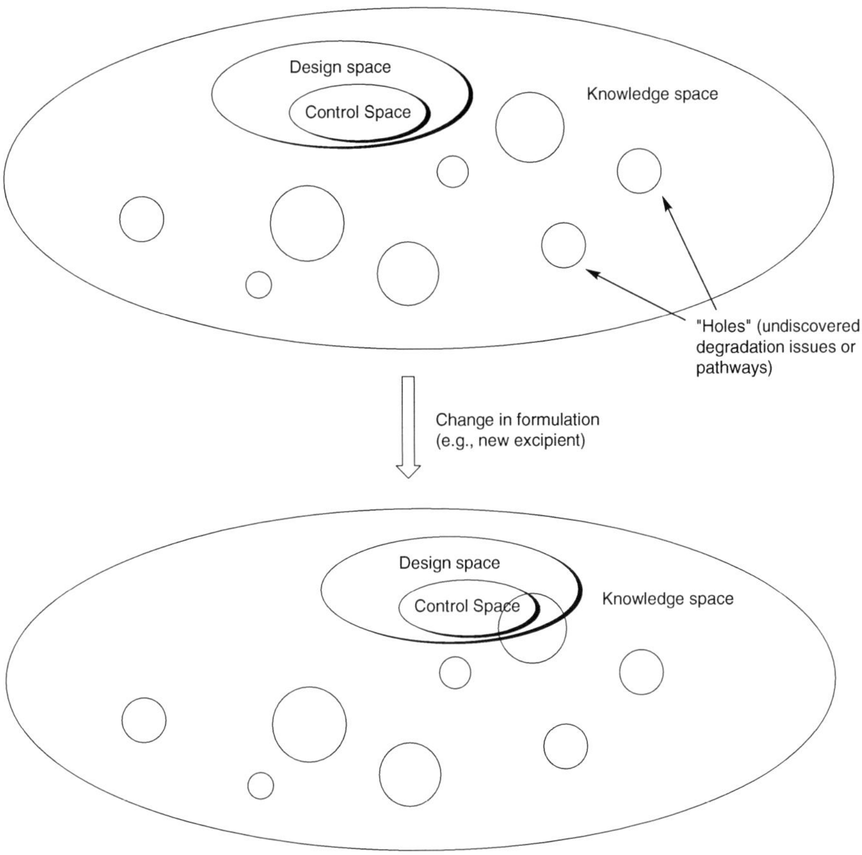

Fig. 15-5. Conceptual representation of a change in the "location" of design and control spaces on the knowledge space surface, resulting in a design and control space that now has "holes," or unexpected/undiscovered degradation-related issues

can result in movement of the "design space" such that previous stress testing studies conducted may not be predictive of all of the stability-related issues. New formulations or excipients can result in new drug-excipient interactions to form novel degradation products requiring new analytical methodology. A list of some of the changes that might induce a "degradation design space" change are:

- Salt form changes.
- Packaging changes.
- Polymorphic form changes.
- New presentations (e.g., solid to liquid formulation).
- Formulation changes within a given presentation.
- New or different analytical technology or methodology.

All of these changes present the possibility that the degradation pathways (or the analytical detection of such pathways) might change, leading to a different degradation profile. Further, these changes also present the possibility that significant differences in the kinetics of degradation might occur, including speeding up, slowing down, or the introduction of non-Arrhenius kinetics (e.g., via formulation or salt form induced solid phase changes or even volatilization of the parent drug (Zannou et al. 2007). Finally, new or different analytical methodologies introduce the possibility that the new degradation products might be discovered or that the measured levels of the products might change. Such risks reinforce the importance of the application of sound scientific and chemistry principles during stress testing and stability investigations.

5. Conclusion

Forced degradation (i.e., stress testing) studies are foundational to develop an understanding of real time stability. Stress testing is used to establish the potential degradation products of a drug, which can be used to develop valid stability-indicating methods. Thus, stress testing is a predictive tool, used to discover the intrinsic stability characteristics (i.e., the degradation chemistry) of a molecule. Poorly designed or incomplete stress testing studies can result in gaps in the degradation chemistry "knowledge space", potentially leading to "stability-indicating" methods that are in fact not stability indicating (i.e., the methods may not detect all relevant degradation products). The use of such methods can lead to undesired and costly surprises during long-term stability studies, potentially affecting all areas of development.

Acknowledgments: Bernard A. Olsen is gratefully acknowledged for helpful discussions and expert review of the content of this chapter.

References

Baertschi SW (2006) Analytical methodologies for discovering and profiling degradation-related impurities. TrAC Trends Analyt Chem 25(8):758–767

Baertschi SW, Nussbaum MA, Jansen PJ (2005) Role of "mass balance" in pharmaceutical stress testing. In: Baertschi SW (ed) Pharmaceutical stress testing: predicting drug degradation, vol 153. Taylor & Francis, New York, pp 181–204

Baertschi SW, Jansen PJ et al. (2006) Presented at the Institute for International Research conference on Forced Degradation, Short Hills, NJ

Boccardi G (2005) Oxidative susceptibility testing. In: Baertschi S (ed) Pharmaceutical stress testing: predicting drug degradation. Taylor & Francis, New York, pp 223–225

Boccardi G, Deleuze C, Gachon M, Palmisano G, Vergnaud JP (1992) Autoxidation of tetrazepam in tablets: prediction of degradation impurities from the oxidative behavior in solution. J Pharm Sci 81(2):183–185

Brustugun J, Tonnesen HH, Klem W, Kjonniksen I (2000) Photosensitized degradation of losartan potassium in an extemporaneous suspension formulation. PDA J Pharm Sci Technol 54:136–143

Guerrieri P, Salameh A, Taylor L (2007) Effect of small levels of impurities on the water vapor sorption behavior of ranitidine HCl. Pharm Res 24(1):147–156

International Conference on Harmonization (1997) Guideline for the photostability testing of new drug substances and products, stability testing of new drug substances and products, Q1B, November 1997

International Conference on Harmonization (2003) Stability testing of new drug substances and products, Q1A (R2), August 2003

International Conference on Harmonization (2006a) Guidance for industry: pharmaceutical development, Q8, May 2006

International Conference on Harmonization (2006b) Pharmaceutical development, Q8 Annex, November 2006

Jansen PJ, Smith KW, Baertschi SW (2005) Stress testing: analytical considerations. In: Baertschi SW (ed) Pharmaceutical stress testing: predicting drug degradation, vol 153. Taylor & Francis, New York, pp 141–171

Nishinaga A, Shinizu T (1980) Base-catalyzed oxygenation of tert-butylated phenols, 4. Mechanism of base-catalyzed ortho regioselective dioxygen incorporation into 4-aryl-2, 6-di-tert-butylphenols. J org chan 45(24): 4997–4998

Olsen BA, Baertschi SW (2003) Strategies for investigation and control of process- and degradation-related impurities. In: Ahuja S, Alsante KM (eds) Handbook of isolation and characterization in pharmaceuticals. Academic, New York, pp 89–117

Reed RA, Harmon P, Manas D, Wasylaschuk W, Galli C, Biddell R, Bergquist PA, Hunke W, Templeton AC, Ip D (2003) The role of excipients and package components in the photostability of liquid formulations. PDA J Pharm Sci Technol 57(5):351–368

Salameh AK, Taylor LS (2005) Deliquescence in binary mixtures. Pharm Res 22(2):318–324

Seburg RA, Ballard JM, Hwang T-L, Sullivan CM (2006) Photosensitized degradation of losartan potassium in an extemporaneous suspension formulation. J Pharm Biomed Anal 42(4):411–422

Thatcher SR, Baertschi SW, Mansfield RK, Miller RB, Davis CW (2001a) Pharmaceutical photostability: a technical guide and practical interpretation of the ICH guideline and its application to pharmaceutical stability – Part 1. Pharm Technol 25(3):98–110

Thatcher SR, Mansfield RK, Davis CW, Baertschi SW (2001b) Pharmaceutical photostability. A technical guide and practical interpretation of the ICH guideline and its application to pharmaceutical stability. Part II. Pharm Technol 25(4):50–62

Wasylaschuk WR, Harmon PA, Wagner G, Harman AB, Templeton AC, Xu H, Reed RA (2007) Evaluation of hydroperoxides in common pharmaceutical excipients. J Pharm Sci 96(1):106–116

Waterman KC, Adami RC, Hong J (2003) Impurities in drug products. In: Ahuja S, Alsante KM (eds) Handbook of isolation and characterization of impurities in pharmaceuticals. Academic, New York, pp 75–88

Zahn M (2006) Temperature control and product quality of medicines in transit, Barcelona, 27–28 Sept 2006, pp 731–736

Zahn M, Kållberg PW, Slappendel GM, Smeenge HM (2006) A risk-based approach to establish stability testing conditions for tropical countries. J Pharm Sci 95(5): 946–965

Zannou EA, Ji Q, Joshi YM, Serajuddin ATM (2007) Stabilization of the maleate salt of a basic drug by adjustment of microenvironmental pH in solid dosage form. Int J Pharm 337(1–2):210–218

Author Biography

Dr. Steve Baertschi is a Research Fellow in Analytical Sciences R&D at Eli Lilly and Company, Indianapolis, IN. He received his B.S. degree in Chemistry from David Lipscomb University (Nashville, TN) in 1980. From 1980 to 1985 he worked in the environmental analytical area focusing on priority pollutant analysis. He obtained his Ph.D. in Organic Chemistry in 1989 from Vanderbilt University (Nashville, TN),and joined Lilly that same year. While at Lilly he has been responsible for developing the areas of stress testing, photostability testing, and impurity identification for small molecule drug candidates. Research interests include impurity isolation/structure elucidation, photostability of drugs, mechanisms of drug degradation, reactive intermediates, and drug-excipient interactions. Dr. Baertschi has organized/chaired numerous scientific conferences and symposia on stress testing, photostability, and impurities; he has published more than 30 scientific articles in refereed journals, 13 book chapters, and edited a book on the topic of pharmaceutical stress testing. Dr. Baertschi has been a member of the American Chemical Society since 1980 and was named a Fellow of the American Association of Pharmaceutical Scientists in 2008.

Chapter 16

Low Level Impurities in Drug Substances and Drug Products and the Analytical Challenges in Identification and Quantitation

Ganapathy Mohan

Abstract Impurities that are seen in the active pharmaceutical ingredients (API) and the degradants that are seen in the drug products are to be identified and quantified according to regulatory requirements, especially when they are recurring ones and above the threshold limits that are described in guidance documents such as the ones from International Conference on harmonization (ICH). The identification and quantification of these impurities become even more important, especially when they could potentially have safety related adverse implications such as genotoxicity. Impurities such as the genotoxic ones and/or others that are potentially unsafe need to be controlled to very low levels or to be eliminated, typically in the order of parts per millions (ppm). This is usually a very challenging endeavor to come up with analytical techniques that have detection systems and methodologies that are very sensitive to support very low levels of detection (LOD) and quantitation (LOQ). This article discusses the implications of such low level impurities in the API and DP and the associated analytical challenges in identifying and detecting these impurities.

1. Introduction

One of the key analytical challenges in drug development is the control of low level impurities. International Conference of Harmonization (ICH) has issued several guidelines to govern this activity, specifically Q3A for Active Pharmaceutical Ingredient (API) and Q3B for drug product. In recent years, analytical technologies have advanced so much that the detection of impurities at very low levels has become significantly less complex though the complexity still poses challenges. There is no current regulatory requirement that dictates the level to which these impurities should be monitored. For an inorganic impurity such as metals, the techniques that are used could easily be detected in the sub-PPM or PPM ranges. But for organic impurities, measuring something to the extent of parts per million can become challenging.

From: *Pharmaceutical Stability Testing to Support Global Markets*: Pharmasp,
Edited by: K. Huynh-Ba, DOI 10.1007/978-1-4419-0889-6_16,

2. Monitoring Low Level Impurities

Regulatory agencies require that the purity of a drug product must be monitored. In addition, the impurities that are above the threshold levels must also be detected and identified. Understanding the formation of genotoxic impurities (GTIs) can also be deciphered based on prior knowledge of the molecules, the manufacturing processes, the starting materials, potential functional groups and so on. Safety is one of the key factors and if not the key factor, that is a driver for monitoring and controlling impurities. These levels are essentially determined by the capability of the methods applied and their sensitivity.

2.1. Typical Techniques that are Used for Identification and Quantification

Ideally, a chromatographic technique is used due to its separation power and the advancement in its detection systems, such as ultraviolet visible (UV) spectroscopy, fluorescence and electrochemical spectroscopy. If higher sensitivity and more universality are needed, other techniques like refractive index, evaporative light scattering detection and mass spectrometry can also be used. Infrared (IR) and nuclear magnetic resonance (NMR) spectroscopy have also been used for identification purposes.

2.2. Impact of Stability

Impurities are actually materials that are present in API, reagents or solvents used in the process. In many cases, these impurities may not be monitored on stability if they do not actually increase with time. *Degradation products or degradants* are monitored on stability. These are products that are formed due to the influence of temperature, humidity, oxidation, exposure to light etc. For the purpose of method development and validation of the analytical methods, possible degradants are created using forced degradation studies by exposing the product to oxygen, heat, light, humidity, acid and base.

However, in the case of low level impurities, specifically the GTIs, these are monitored as appropriate, typically in the API.

Identification and detection of these impurities could depend on the timing and the phase of the development. The challenges of the low level impurities measurements pose needs to balance efforts with considerations of the stage of development, potential for changing the synthetic route, impact on patient safety and constraints on time and resources involved. There are a couple of concepts for achieving this. They are As Low As Reasonably Achievable (ALARA), and As Low As Reasonably Practicable (ALARP). The ALARA approach requires more time and money, as this would mean that methods have to be developed which can detect as low as we can. ALARP option is a more practical approach as one could use all the balancing factors mentioned above in developing methods that are more pragmatic and practical for the case in point.

3. Genotoxic Impurities

The synthesis of pharmaceuticals can lead to the production of GTIs. Due to the possible carcinogenic activity, these impurities may potentially cause different dose response curves, and thus require limits that are other than the non-qualified impurities. Unfortunately, there was no written guidance document from FDA for controlling this category of impurities until 2009; although there is a guidance from the European Authorities.

The levels of these impurities play a critical role in the drug development process. One of the potential impacts on development is a clinical hold and therefore, may delay the development of a drug candidate in addition to posing significant analytical chemistry challenges.

3.1. Key Principles

Acceptable levels are based on the maturity of the synthetic scheme, the availability of analytical methods, the duration or phase of clinical trials, number of participants, type of patients, indication, and other potential risks. There are various strategies proposed to determine Acceptable Levels. In some cases, 1 PPM can be considered the limit of GTIs; since several companies in the early days of GTI monitoring awareness (in the mid 2000s) received clinical holds with GTI levels above 1 PPM, which resulted in delaying of the development of those drugs. Recently, a level of approximately 1.5 μg/day was introduced using the concept of threshold of toxicological concern (TTC) for GTIs.

3.2. PhRMA Position Paper and *Staged TTC Approach*

PhRMA Task Force published a paper in Regulatory Toxicology and Pharmacology in 2006, where in the acceptable limits in μg/day was calculated based on allowable daily intake calculations and using a staged TTC approach, as well as the calculated risk assessment (see Table 16-1). This publication gives a procedure for testing, classification, determining toxicological risk assessment of GTIs in pharmaceutical products. Its recommendations cover all phases of development for new drug applications. In addition, it provides a mechanism for taking the phase of development and therefore the duration of clinical trials in the overall strategy of GTI control.

3.3. Proposed Qualification of Impurities

The PhRMA publication recommends three main steps:

1. Identify and classify structural alerts in your parent compounds and impurities.
2. Establish a qualification strategy based upon that classification.
3. Establish acceptable limits of this impurity in API based upon the allowable daily intake and the duration of exposure to patients (i.e., clinical trials)

This approach does not strictly propose a value, but it considers the allowable daily intake and the duration of the clinical trials; therefore, this approach provides a more complete method to monitor GTIs when progressing along the development cycle.

Table 16-1. Acceptable limits in micrograms per day based on allowable daily intake calculations using a staged TTC approach.

	Duration of Exposure				
	≤1 mo.	>1-3 mo.	>3-6 mo.	>6-12 mo.	>12 mo.
Proposed Allowable Daily Intake (μg/day) for all Phases of development	**120** or	**40** or	**20** or	**10** or	**1.5**
Alternative maximum based on percentage of impurity in API	0.5%	0.5%	0.5%	0.5%	

10^{-6} cancer risk–extra conservatism during shorter duration trials (e.g. for volunteers)
10^{-5} cancer risk– risk used by ICH for carcinogenic residual solvents and CHMP draft for genotoxic impurities

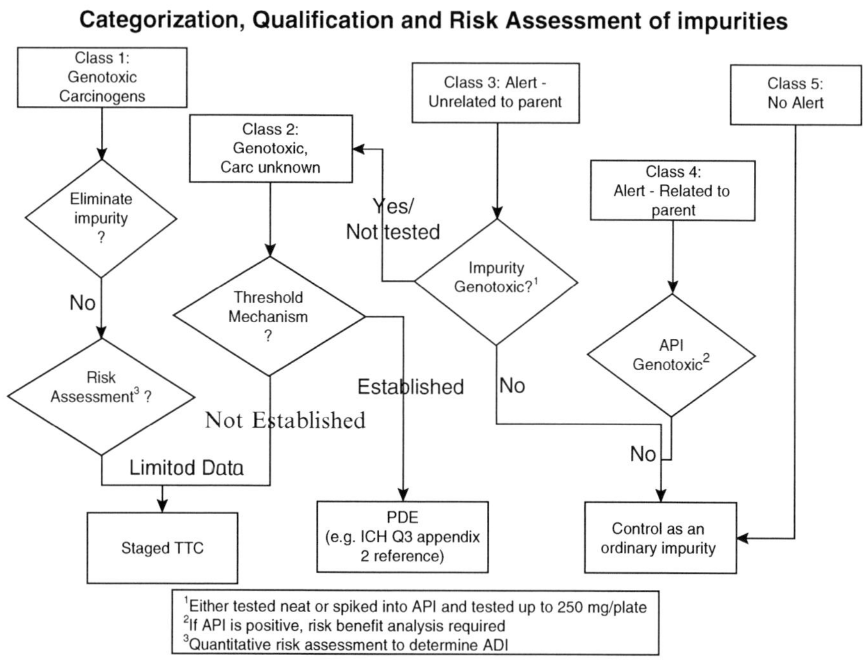

Fig. 16-1. Categorization, qualification and risk assessment of impurities

3.4. Scope of EMEA Guideline

European Medicines Evaluation Agency (EMEA) issued a guideline in June 2006 recommending that GTIs in new API be monitored. At this time, this guideline does not apply to approved products except new applications where there are variations to the marketing applications. When changes to the synthesis or manufacturing processes occur, work must be done to ensure that no new or higher levels of GTIs are introduced. It should also be evaluated against the database of existing marketed products of the same drug products.

It also focuses on DNA reactive substances with potential for direct DNA damage and non-threshold related mechanism in contrast to compounds interacting with non-DNA targets. It applies only to new drug application unless there is a specific concern of an approved product, or to specific application during clinical development phases.

4. Conclusion

GTIs that are also low-level impurities need to be monitored like any other impurities, especially when they are recurring impurities, and it is important to understand the source of these. FDA has not issued a guidance document at the time of this presentation. This document and the EU guidance document will help pharmaceutical industry to develop consistent approaches to manage the GTIs in API, keeping in mind that patient safety is the first and foremost factor. Analytical methods need to be evaluated constantly so as to monitor these impurities, and this provides a challenge that becomes achievable with the advancement in the technology.

References

"PhRMA Genetoxic Impurities-Task Force Publication", PhRMA Task Force, Regulatory Toxicology and Pharmacology, 2006, 44: 198-211.

Author Biography

Dr. Ganapathy Mohan has a Ph.D. in Analytical Chemistry from Kansas State University. He has 25 years of experience in Pharmaceutical Industry. Mohan is currently Executive Director at Merck and Company and heads the Global Pharmaceutical CMC department and Labeling Services.

Prior to this, Mohan worked at Sanofi-aventis Pharmaceuticals where he had held many responsibilities in Analytical Sciences, Quality Control, CMC and Quality Assurance departments. His last position in Sanofi-aventis was Associate Vice President of Global Analytical Sciences Department. He is an active member of AAPS and PhRMA and was a member of the Editorial Board for the Journal of PAT.

Chapter 17

Stability of Repackaged Products*

Mansoor A. Khan

Abstract For better patient compliance and easier distribution, some drug products are frequently repackaged from their original containers into unit-dose packages by pharmacists or commercial repackagers. The USP indicates that for non-sterile solid dosage forms packaged in unit-dose containers, the beyond-use date shall be 1 year from the date the drug is packaged into the unit-dose container or the expiration date on the manufacturer's container, whichever is earlier, unless stability data or the manufacturer's labeling indicates otherwise. Due to differences in moisture permeation, thermal susceptibility or surface chemistry of the new packaging material when compared to the original, the stability of a repackaged drug product can differ from that obtained from original container. This concern prompted us to study the stability issues associated with certain repackaged products. This chapter will highlight some of the findings in our research laboratory.

1. Background

Repackaging of drug products into unit-dose packaging systems has found greater use in the healthcare. Repackaged products are available either directly from manufacturers and repackagers, or from hospitals and community pharmacies. In this regard, often the packaging is changed from their original container which is different than the container in which the stability of the drug product was originally established. Thus, there is an ambiguity about assigning an appropriate expiration dating to repackaged drug products in the absence of stability testing.

A guidance from Food and Drug Administration (FDA) states that no action will be initiated against any unit dose repackaging firm or drug product in a unit dose container solely on the basis of the failure of the repackaging firm to have stability studies supporting the expiration dates used if they are meeting all

*Disclaimer: The views expressed in this book chapter are only of the author and do not necessarily reflect the policy of the agency.

From: *Pharmaceutical Stability Testing to Support Global Markets*: Pharmasp,
Edited by: K. Huynh-Ba, DOI 10.1007/978-1-4419-0889-6_17,

other conditions of FDA's repackaging requirements (FDA/ORA 1995). These conditions include, but are not limited to: (1) the unit dose container complies with the class A or class B United States Pharmacopeia (USP) standard, (2) the expiration period does not exceed 6 months after repackaging, (3) the 6 month expiration period does not exceed 25% of the remaining time between the date of repackaging and the expiration date on the original manufacturer's bulk container, and (4) the bulk container has not been previously opened. In May 2005 FDA published a draft guidance titled "Expiration Dating of Unit-dose Repackaged Drugs: Compliance Policy Guide" proposing a revision of the previous policy (FDA guidance 2005). This draft guidance states that the FDA will not initiate action against a non-sterile, unit dose repackaging firm or drug product in unit dose container solely on the basis of the failure of the repackaging firm to have stability studies supporting the expiration dates used, provided the repackager meets all other regulations applicable to repackaged drug products and complies with certain recommendations. These recommendations include, but are not limited to: (a) the expiration date does not exceed 1 year from the date of repackaging or (b) does not exceed the expiration date on the container of the original manufacturer's product, whichever is earlier, unless stability data or the original manufacturer's product labeling indicates otherwise.

However, there seem to be an increasing concern regarding quality of repackaged drug products. In a Technical Report Series, WHO has raised concerns about the potential for increased risk of contamination, cross-contamination, mix-ups, degradation, and changes in physical properties as a result of repackaging drugs (WHO 2003). Review of the scientific literature indicates there are relatively few laboratory studies that evaluate the effect of repackaging on drug products. Moisture permeation, weight gain resulting from moisture ingress into repackaged products, and changes in the stability profile of a drug product if it is sensitive to moisture content are some of the concerns for repackaged drug products (Reamer and Grady 1997).

FDA initiated studies to determine the stability and quality of repackaged drug products to address some of these issues. Five drug products which were evaluated for studying the stability of repackaged products included ranitidine hydrochloride syrup, phenytoin suspension, gabapentin capsule, furosemide tablet, and metoprolol tartrate tablet. Their structure, formulation type, and strength of the drug products which were evaluated are listed in Table 17-1. The comparisons were made with the drug products in their original (non-repackaged) containers.

2. Ranitidine Hydrochloride Syrup

Ranitidine syrup was chosen for the study to evaluate stability as it is very highly used over-the-counter medicine for the treatment of heart burn and Zollinger–Ellison syndrome for pediatric patients. The original and repackaged syrup containers were exposed to International Committee of Harmonization (ICH) conditions for aqueous drug products at accelerated (40°C/25% relative humidity (RH), 3 months) and long-term (25°C/40% RH, 12 months) conditions (ICHQ1AR2 2005). At pre-determined time intervals, the samples were evaluated for pH, weight loss and potency as well as appearance and quantification

Table 17-1. Drug products used in repackaging study.

Name	Structure	Formulation type	Strength	Unit-dose packaging material/amount
Ranitidine hydrochloride	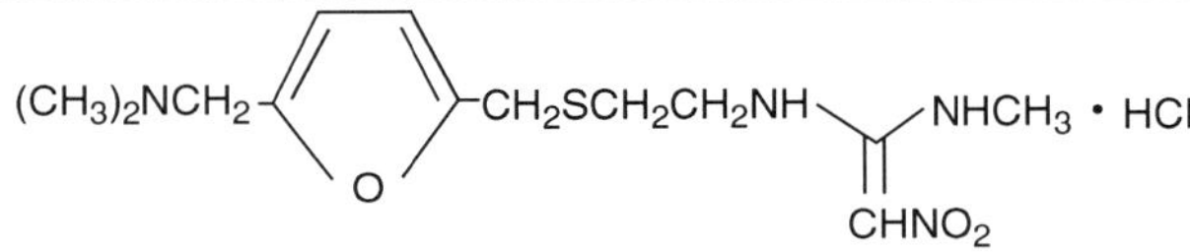	Syrup	15 mg/mL	Amber glass bottle with child-resistant cap/5 mL
Phenytoin sodium	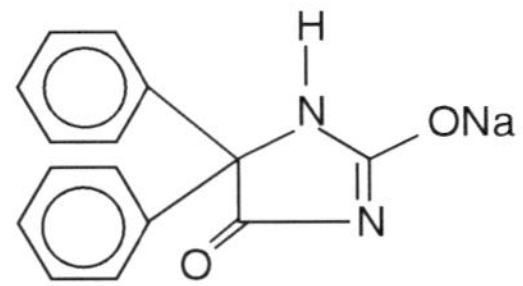	Suspension	25 mg/mL	Amber glass bottle with child-resistant cap/8 mL
Gabapentin	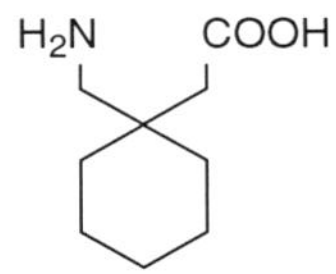	Capsule	300 mg	Unit dose blister strip/ 1 capsule
Furosemide	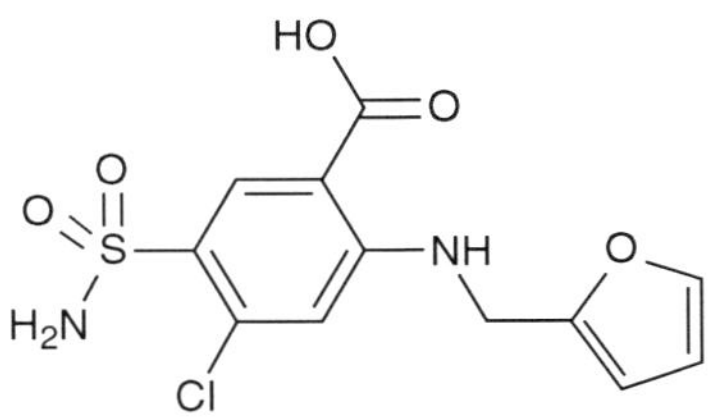	Tablet	40 mg	Blister package/1 tablet
Metoprolol tartrate	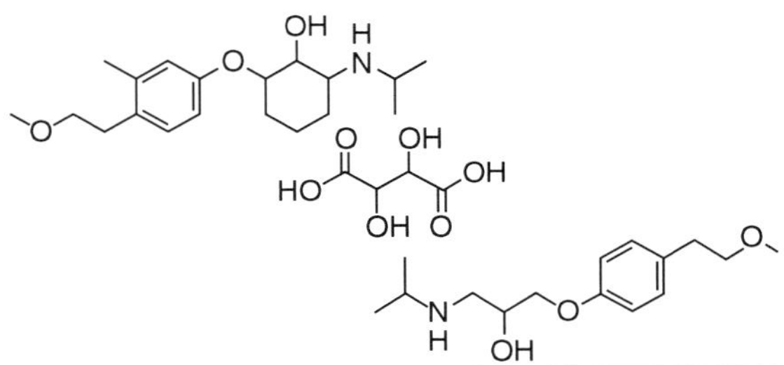	Tablet	50 mg	Blister package/1 tablet

Table 17-2. Ranitidine assay and major impurity related compound C formation for original and repackaged syrup drug products under long-term and accelerated stability conditions.

Exposure	Sample	4 weeks	9 weeks	13 weeks	26 weeks	39 weeks	52 weeks
Ranitidine assay (%)							
25°C/40% RH	Original	98.07 ± 0.53	–	94.2 ± 0.085	103.3 ± 0.417	101.2 ± 0.46	101.7 ± 0.29
	Repackaged	98.23 ± 0.46	–	95.25 ± 0.57	104.5 ± 1.16	99.62 ± 4.97	104.06 ± 0.60
40°C/25% RH	Original	97.66 ± 0.28	99.64 ± 0.85	93.2 ± 0.16	–	–	–
	Repackaged	100.7 ± 0.42	104.4 ± 0.85	101.45 ± 0.17	–	–	–
Related impurity C (%)							
25°C/40% RH	Original	0	–	0.21 ± 0.002	0.30 ± 0.002	0.44 ± 0.008	0.53 ± 0.009
	Repackaged	0	–	0.21 ± 0.001	0.31 ± 0.00195	0.45 ± 0.005	0.53 ± 0.009
40°C/25% RH	Original	0.4 ± 0.005	0.83 ± 0.009	1.03 ± 0.0015	–	–	–
	Repackaged	0.87 ± 0.58	0.79 ± 0.005	1.008 ± 0.013	–	–	–

of impurities. A validated HPLC method published by Shah et al. was used to assay the drug and impurities (Shah et al. 2006).

It was observed that the appearance of impurities was higher in repackaged unit-dose container at the end of accelerated stability testing. However it was within acceptable limits for impurities according to USP guidelines. The drug assay and impurity profiles were well within the acceptable limits under both stability conditions (Table 17-2). The pH and weight change were also not significantly affected. Thus, the unit-dose containers of ranitidine syrup were stable throughout the study period of 3 and 12 months under accelerated and long-term stability conditions (Shah et al. 2008). However, the authors cautioned that the results of the study should not be generalized for any other syrup product or for ranitidine syrup in any other types of containers. A systematic study was warranted on a case by case basis to determine the shelf-life of repackaged drug products.

3. Phenytoin Sodium Suspension

Phenytoin is an anticonvulsant drug which can be useful in the treatment of epilepsy for children experiencing tonic-clonic (grand mal) seizures in suspension form. Repackaging a suspension from its original container would require thorough shaking and redistribution of the settled particles prior to dispensing in the repackaged containers.More often, repackaging is not done by the manufacturing firms directly and therefore, a suspension drug product was included in the study to determine its stability.

The original and repackaged suspension containers were exposed to ICH conditions for aqueous drug products (ICHQ1AR2 2005). At pre-determined

Table 17-3. Phenytoin assay and dissolution values for original and repackaged suspension drug products under long-term and accelerated stability conditions.

Exposure	Sample	4 weeks	8 weeks	13 weeks	26 weeks	39 weeks	52 weeks
Phenytoin assay (%)							
25°C/40% RH	Original	91.08 ± 12.46	–	97.35 ± 8.61	104.32 ± 1.15	98.66 ± 4.26	84.71 ± 22.17
	Repackaged	99.66 ± 1.26	–	101.28 ± 3.81	95.19 ± 7.98	105.32 ± 1.64	101.30 ± 3.33
40°C/25% RH	Original	86.43 ± 12.46	98.68 ± 10.43	94.09 ± 16.20	–	–	–
	Repackaged	93.51 ± 5.99	91.63 ± 10.30	101.04 ± 8.79	–	–	–
Phenytoin dissolution (%)							
25°C/40% RH	Original	96.44 ± 6.53	–	95.89 ± 6.97	94.11 ± 5.95	101.33 ± 8.56	100.41 ± 3.93
	Repackaged	93.82 ± 7.45	–	93.65 ± 4.37	95.52 ± 7.16	102.33 ± 3.55	103.37 ± 2.86
40°C/25% RH	Original	101.02 ± 4.42	94.97 ± 7.71	98.85 ± 5.55	–	–	–
	Repackaged	99.91 ± 3.76	93.52 ± 5.55	99.20 ± 3.06	–	–	–

time intervals, the samples were evaluated for pH, weight loss, potency, and dissolution.

The measured percentage release of phenytoin in both original and repackaged containers under long-term and accelerated conditions met the dissolution specifications. However, assay results showed that the specifications were not met for original as well as repackaged samples under accelerated stability conditions (Table 17-3). The potency samples under long-term condition exhibited more failures for original container as compared to repackaged samples.

The possible explanation of the results might be differences in the technique with which the manufacturer assays the drug from the suspension product as compared to the lab scientists in the current study. Also, the original product in narrow-mouth containers might not provide adequate airspace above the liquid for proper shaking. Chemistry-Manufacturing-Control aspects might need further evaluation for understanding the reasons for product failure in their original containers.

4. Gabapentin Capsules

Gabapentin is a γ-aminobutyric acid (GABA) analog used for the treatment of seizures in adults and children. Gabapentin degrades via intramolecular cyclization to form a γ-lactam which has been shown to cause seizures in an animal model (Potschka et al. 2000). Due to this potentially harmful degradation product and gabapentin's widespread use, gabapentin was chosen as a model drug to test the draft guidance expiration limit of 12 months for drug product capsules following repackaging.

The original and repackaged syrup containers were exposed to ICH conditions for solid dosage forms at accelerated (40°C/75% RH, 3 months) and long-term (25°C/60% RH, 12 months) conditions (ICHQ1AR2 2005). At pre-determined time intervals, the samples were evaluated for weight loss,

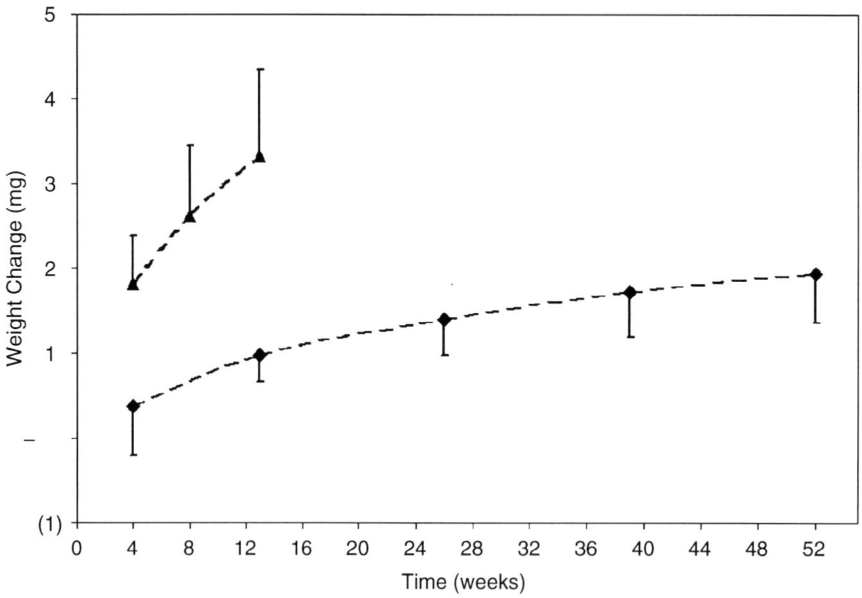

Fig. 17-1. Weight change of the unit-dose repackaged blister strips of gabapentin capsules stored under long-term (*filled diamond*) and accelerated (*filled triangle*) storage conditions

Table 17-4. Gabapentin assay for original and repackaged capsule drug products under long-term and accelerated stability conditions.

Exposure	Sample	4 weeks	8 weeks	13 weeks	25 weeks	39 weeks	52 weeks
25°C/40% RH	Original	97.5±0.2	–	97.1±2.4	99.6±0.3	97.9±0.4	98.9±0.2
	Repackaged	97.5±0.3	–	99.5±1.6	99.1±1.1	97.9±0.2	98.8±0.2
40°C/25% RH	Original	95.6±0.8	96.5±2.1	98.6±0.7	–	–	–
	Repackaged	95.7±0.4	97.7±0.2	97.2±4.0	–	–	–

potency as well as appearance and quantification of impurities, and dissolution studies. Previously published validated HPLC method was used to assay the drug and impurities (Ciavarella et al. 2007; Gupta et al. 2008a).

The unopened original containers of the drug product showed no weight change during the 13 and 52 weeks under the accelerated and the long-term storage conditions, respectively. However, the unopened individual units of the repackaged drug product under both storage conditions showed small, though significant, weight increase (Fig. 17-1). The potency results showed no difference between the original and repackaged drug products (Table 17-4). Samples were also analyzed for gabapentin's main degradation product, the lactam. Only, capsules stored under accelerated storage conditions for 13 weeks showed quantifiable levels (≥0.1%) of lactam degradation product. The lactam levels were slightly higher for repackaged drug products (0.17%) as compared to original (0.11%) but they were well below the acceptable limit of 0.4%. No differences were observed between the dissolution profiles for all samples (Gupta et al., 2008b).

Table 17-5. Furosemide assay and water loss for original and repackaged tablet drug products under long-term and accelerated stability conditions.

Exposure	Sample	4 weeks	8 weeks	13 weeks	25 weeks	39 weeks	52 weeks
Furosemide assay (%)							
25°C/40% RH	Original	94.6±2.0	–	92.3±0.9	91.5±0.9	94.8±2.2	92.5±3.3
	Repackaged	91.7±1.0	–	91.7±1.5	91.7±1.2	95.9±3.8	91.6±4.0
40°C/25% RH	Original	98.3±1.2	100.3±2.9	92.3±0.9	–	–	–
	Repackaged	95.1±3.8	94.7±3.6	91.7±1.5	–	–	–
Water loss (%)							
Exposure	Sample	4 weeks	8 weeks	13 weeks	25 weeks	39 weeks	52 weeks
25°C/40% RH	Original	3.2	–	3.95	3.15	2.35	3.15
	Repackaged	3.2	–	3.15	2.35	2.35	3.15
40°C/25% RH	Original	3.15	2.4	3.15	–	–	–
	Repackaged	2.4	3.2	2.4	–	–	–

It was concluded that the gabapentin capsules tested in the study were stable up to a period of 1 year under long-term storage conditions and 3 months under accelerated storage conditions in original and repackaged containers. However, authors believed that the results of this study should not be extrapolated to other drug products.

5. Furosemide Tablets

Furosemide is a highly prescribed powerful diuretic that is used to treat excessive accumulation of fluid and/or edema of the body caused by heart failure, cirrhosis, chronic kidney failure, and the nephrotic syndrome. It is sometimes used alone or in conjunction with other blood pressure pills to treat high blood pressure.

The stability of 40 mg furosemide tablet product repackaged in a unit-dose USP class A blister packs and identical product in its original containers made of HDPF material was studied under ICH conditions for solid dosage forms (ICHQ1AR2 2005). Samples that were periodically removed from the two storage conditions were subjected to these tests: (1) water loss on heating, (2) potency assay, (3) tablet hardness, (4) dissolution, (5) TGA/DSC (6) spectroscopic – NIR, Raman, FTIR, NIR imaging – analyses.

The results from these tests indicated that there were no differences in the stability of the repackaged product when compared to the original product in its original package during 12 months study period (Table 17-5). Tablet hardness was about 6.0±0.3 KP for both the original and the repackaged products under both long-term and stressed conditions. The mean percentage of water loss on heating are given in Table 17-5. The TGA/DSC and the spectroscopic data did not show any discernible differences in the products.

Based on the results of this study, it was concluded that under the conditions of the study, there were no observable differences in both the original and repackaged products, and that stressing them did not produce any measurable differences. Thus product quality attributes of the furosemide tablets were not affected by the repackaging.

6. Metoprolol Tartrate Tablets

Metoprolol tartrate tablets are used for treating high blood pressure, alone or with other medicines; long-term treatment of chest pain; and reducing the risk of death because of heart problems in patients who have had a heart attack. This is one of the most prescribed beta blockers for maintenance of blood pressure or to stabilize the heart rhythm. This formulation was chosen to study stability of the drug product in the unit-dose Class A blister package as it contained excipients which had higher potential of moisture uptake. The inactive ingredients listed in the package insert included povidone, microcrystalline cellulose, hydroxypropyl methyl cellulose, sodium starch glycolate, polysorbate, and colloidal silicon dioxide.

Stability studies were conducted as described in gapapentin capsules section. The drug products were analyzed for potency, dissolution, water content, loss on drying and hardness. Near infrared (NIR) chemical imaging was also used to detect changes in the drug product during storage under stability conditions.

The results indicated no differences in the stability of the repackaged tablets as compared to the tablets from the original container during the 12 months storage under long-term condition (Table 17-6). However, a significant increase in weight due to moisture uptake was observed for the repackaged tablets stored under accelerated conditions over the 13-week study period, while the tablets stored in the original container remained unchanged. The increase in tablet weight was accompanied by a decrease in tablet hardness from 8 kp to ~0 kp and a significant increase in dissolution rate from 51 to 92% of drug released in 5 min (Fig. 17-2). NIR chemical images of the repackaged tablets stored under accelerated condition were also found to be significantly different from the images collected on the original tablets (Fig. 17-3).

Table 17-6. Metoprolol assay, hardness, and water loss for original and repackaged tablet drug products under long-term and accelerated stability conditions.

Exposure	Sample	4 weeks	8 weeks	13 weeks	25 weeks	39 weeks	52 weeks
Metoprolol assay (%)							
25°C/40% RH	Original	99.4±0.1	–	100.9±1.1	98.1±0.5	96.6±1.0	98.7±1.0
	Repackaged	100.4±1.2	–	100.2±1.0	100.2±0.8	97.8±2.2	99.4±1.0
40°C/25% RH	Original	100.6±0.6	97.2±1.3	92.3±0.9	–	–	–
	Repackaged	104.9±0.6	101.9±1.2	102.42±1.6	–	–	–
Hardness (kP)							
25°C/40% RH	Original	8.2±1.1	–	8.2±0.6	8.7±1.2	8.9±1.1	8.2±1.1
	Repackaged	7.4±1.4	–	7.7±1.5	7.9±0.7	7.9±0.8	7.4±1.4
40°C/25% RH	Original	8.4±1.2	8.4±0.8	8.0±0.7	–	–	–
	Repackaged	0	0	0	–	–	–
Water loss (%)							
25°C/40% RH	Original	3.55	–	2.85	3.55	3.55	3.55
	Repackaged	4.25	–	4.60	4.95	4.30	4.30
40°C/25% RH	Original	3.65	3.55	4.30	–	–	–
	Repackaged	9.40	9.45	10.65	–	–	–

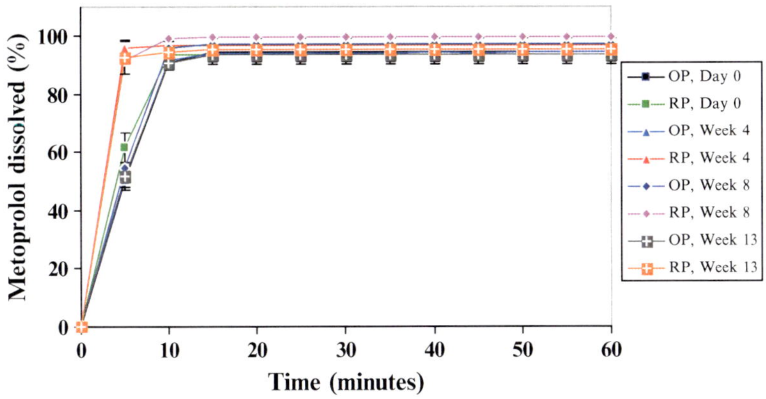

Fig. 17-2. Dissolution profiles of original and repackaged metoprolol tartrate tablet at 40°C/75% RH

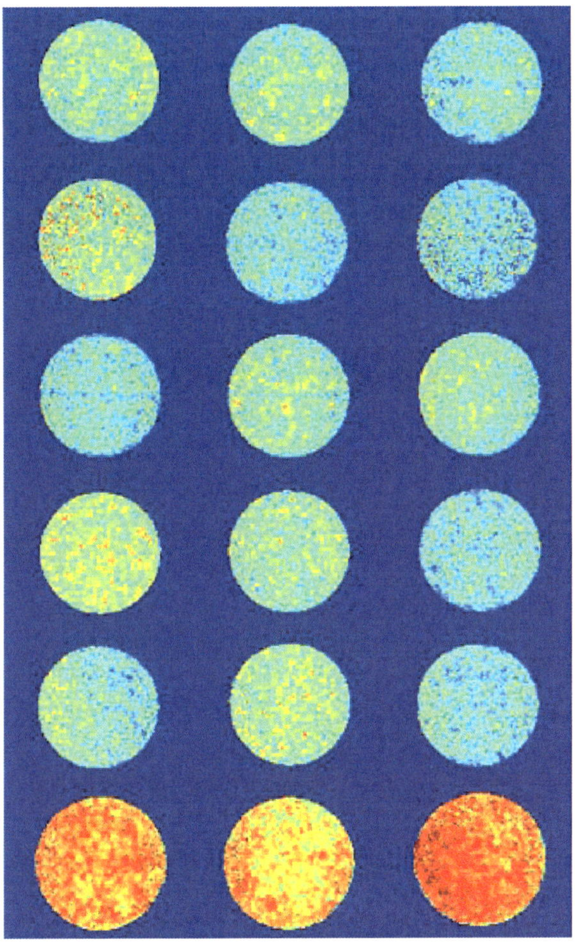

Fig. 17-3. NIR chemical images of the tablets from the original containers (OP) and repackaged unit-dose blister strips (RP) at the start of the study (0 weeks) and after 52 and 13 weeks storage at 25°C/60% RH and 40°C/75% RH, respectively

The observed change in the product quality of the repackaged metoprolol tablets may adversely affect the products bioavailability profile making it unsuitable for patient administration, even though the potency of the active drug remained within the USP range of 90–110%. The moisture taken resulted in: (1) an increase in tablet weight, (2) a decrease in tablet hardness and (3) faster tablet dissolution.

7. Conclusions

Lack of stability problems for repackaged ranitidine syrup, gabapentin capsules, and furosemide tablets cannot be generalized to other drug products. As from the research studies in a FDA lab, it was found that the product quality of metoprolol tartrate tablets in repackaged pack was compromised due to permeation of moisture across the USP Class A materials. The study warrants that USP Class B materials with significantly higher moisture permeation should be avoided for repackaging drug products, especially if they are formulated using hygroscopic ingredients. For suspension dosage form, redispersibility of the cake could become a potential problem in the failure of assay. The study highlights the importance of selecting the packaging material for repackaging a drug product based on the scientific understanding of the hygroscopicity of different formulation components, in addition to the active drug substance. For these products applying the same expiration to the repackaged drug products without any stability studies could compromise their product quality during storage even when repackaged using USP Class A materials. Stability failure is a significant safety concern which could not only result in their sub-optimal efficacy but also potential toxic impurity generation. Gabapentin lactam formation and increase in ranitidine related compound C are examples of this although from repackaged container, the levels were well below the permissible limits. However, it cannot be ruled out that this might become a potential safety concern for other drug products. Repackaging is desirable by pharmacies, hospitals and from patients' point of view. However, science- and risk-based decisions are necessary for assigning an appropriate expiry date for the repackaged drug products.

Acknowledgments Authors would like to thank Barry Rothman from Office of Compliance and DPQR scientists: Agnes Nguyenpho, Robert Hunt, Hulahalli Prasanna, Rakhi Shah, Ebenezer Asafu-Adjaye, Arthur Bryant, Yongsheng Yang, Alan Carlin, Abhay Gupta, Tony Ciavarella, Everett Jefferson, Patrick Faustino, Donna Volpe, Susan Jenney, Chris Ellison, Mazen Hamad, and Mobin Tawakkul.

Reference

Ciavarella AB, Gupta A, Sayeed VA, Khan MA, Faustino PJ (2007) Development and application of a validated HPLC method for the determination of gabapentin and its major degradation impurity in drug products. J Pharm Biomed Anal 43:1647–1653

FDA draft guidance for Industry (2005) Expiration dating of unit-dose repackaged drugs: compliance policy guide. US Food and Drug Administration, Rockville, MD, May. http://www.fda.gov/cder/guidance/6169dft.htm.

FDA/ORA Compliance Policy Guide 7132b.11 (1995) Expiration dating of unit-dose repackaged drugs. US Food and Drug Administration, Rockville, MD,

March. Available at http://www.fda.gov/ora/compliance_ref/cpg/cpgdrg/cpg480-200.html.
Gupta AB, Ciavarella A, Rothman B, Faustino PJ, Khan MA (2009) Stability of gabapentin 300-mg capsules repackaged in unit dose containers. Am J Health Syst Pharm 66:1376–1380
Gupta AB, Ciavarella A, Sayeed VA, Khan MA, Faustino PJ (2008a) Development and application of a validated HPLC method for the analysis of dissolution samples of gabapentin drug products. J Pharm Biomed Anal 46:181–186
ICHQ1AR2 (2005) Stability testing of new drug substances and products. International Conference on Harmonization, Geneva, Switzerland, November
Potschka H, Feuerstein TJ, Loscher W (2000) Gabapentin–lactam, a close analogue of the anticonvulsant gabapentin, exerts convulsant activity in amygdala kindled rats. Arch Pharmacol 361:200–205
Reamer JT, Grady LT (1997) Moisture permeation of newer unit-dose repackaging materials. Am J Hosp Pharm 35:787–793
Shah RB, Prasanna HR, Rothman B, Khan MA (2008) Stability of ranitidine syrup in repackaged unit-dose container. Am J Health Syst Pharm 65:325–329
Shah RB, Tawakkul MA, Prasanna HR, Faustino PJ, Nguyenpho A, Khan MA (2006) Development of a validated stability indicating HPLC method for ranitidine hydrochloride syrup. Clin Res Regul Aff 23:35–51
World Health Organization (2003) Repackaging and relabelling. In: Good trade and distribution practices for pharmaceutical starting materials. WHO Technical Report Series No. 917, Annex 2, 49–50

Author Biography

Dr. Mansoor A. Khan is the Director of Division of Product Quality Research in Center for Drug Evaluation and Research at US Food and Drug Administration. Prior to joining FDA, Dr. Khan was a Professor of Pharmaceutics and Director of Graduate Program in the School of Pharmacy at Texas Tech University Health Sciences Center. He is a registered pharmacist, and has earned his Ph.D. degree in Industrial Pharmacy from the St. John's University School of Pharmacy at New York in 1992. He has published over 160 peer-reviewed manuscripts, four texts including the "Pharmaceutical and Clinical Calculations," 8 book chapters, and more than 125 poster presentations and over 100 invited oral presentations in various meetings. Dr. Khan graduated 10 Ph.Ds in pharmaceutics. He is the Chair-elect of Formulation Design and Delivery section of AAPS, and has been recognized as AAPS Fellow. He serves on the editorial board of Pharmaceutical Technology and the Journal of Clinical Research and Regulatory Affairs.

Chapter 18

Packaging-Induced Interactions and Degradation

Mark D. Argentine and Patrick J. Jansen

Abstract Forced degradation studies are typically used to determine the potential degradation products that may form during formal stability studies that are conducted on drug substances and drug products. Occasionally, impurities are detected in stability studies that were not observed in the forced degradation studies. Such observations can result from poorly designed forced degradation studies (of the drug substance and/or drug product) that failed to generate all of the relevant degradation products. Potential interaction of the drug with packaging or shipping materials may not be effectively captured unless studies are designed to look for these potential interactions. Migration of materials into drug substance or product may result in simple contamination. However, migratory species may also be reactive and form new drug-related impurities. Several examples are shared to highlight drug substance and drug product interactions with packaging components that fit into the above categories.

1. Introduction

During pharmaceutical development and manufacturing, packaging plays an important role in the protection of product material for ensuring delivery of a safe and efficacious drug material to the patient. Since the primary role of packaging is to protect a drug substance or drug product from its surroundings, it is often considered an inert material (particularly the glass ampoules for injectable products and plastic bags for drug substance and bulk oral product materials), and packaging-related interactions might be initially overlooked as an important component in the overall control strategy for both product safety and elegance. Packaging-related interactions between the drug substances and products have had plenty of historical precedent. Summaries have captured many packaging-related interactions over the past 40 years (Jaminet 1968; Takacsi 1989; Pellerin et al. 1991; Jenke 2002; Fliszar et al. 2006; Balough et al. 2007), including reactivity with stoppers and plastic liners in addition to the extractables and leachables that often challenge impurity profiles for parenteral drug product materials.

From: *Pharmaceutical Stability Testing to Support Global Markets*: Pharmasp,
Edited by: K. Huynh-Ba, DOI 10.1007/978-1-4419-0889-6_18,

The intent of this article is to highlight the continued need for focus on drug-packaging related interactions as a potential source for impurity formation. Packaging-related interactions can take several forms. A few are listed below:

1. Contamination resulting from migration of a packaging component.
2. Interaction of the drug substance or product with a migratory packaging component.
3. Reactivity of the drug with the packaging material itself.
4. Reactivity of the drug material based upon packaging defects or impurities.

Some of these interactions have the potential to directly impact drug safety with the formation and introduction of new impurities into the delivered drug product. Others have the potential to cause elegance concerns by causing product or packaging discolorations that question the integrity of each material.

The first few examples will highlight selected interactions between a drug substance and its packaging material. Therefore, it is helpful to provide a brief background regarding packaging-related requirements and practices for drug substance and product material. One primary attribute of a desirable package include the use of packaging materials that are compatible with the drug and its safety . As the primary packaging material comes into direct contact with the drug, the liner should meet regulatory expectations for food-grade packaging for orally administered drugs to ensure their safety.[1] Additional regulatory expectaions exist in the drugs used for parental formulations.[2] Low-density polyethylene (LDPE)-based liners are often used because of their chemical compatibility with a wide array of materials. Additional packaging items might be added for safety reasons or additional product protection.[3] As an example, an antistatic primary liner is desired for safety considerations and is required in certain countries for packaging in solvent-containing environments (very typical for small molecule drug substance isolations where product is dried and then loaded into storage drums). The conductive liner allows dissipation of static electricity that might build as solid powders flow from dryers to drums and avoids a static discharge powder explosion. The conductive nature is achieved through the use of packaging additives to the plastic liner itself, and several additives have been assessed for safety at low levels. Ensuring compliance with the appropriate indirect food additive regulation (CFR) is considered appropriate to establish safety when packaging additives are used. Secondary packaging is occasionally employed to provide enhanced product protection. For example, laminated foil liners might be used to provide additional water, air (oxygen), or light protection.

In addition to ensuring packaging-related compliance to the above safety-related regulations, material compatibility assessments are important additional safety-related considerations. Stability studies that use packaged material are generally employed to establish suitability of the packaging material with the

[1] As example, see 21 CFR parts 174–189. Specifically, refer to 21 CFR part 177 for low density polyethylene (LDPE)-based food-packaging materials.

[2] As examples, see 21 CFR 310.509, JP XV (General tests, 51. Test Methods for Plastic Containers) and PhEur 3.1 (Materials used for the Manufacture of Containers… section 3.1.4 Polyethylene without additives for containers for parenteral and ophthalmic preparations).

[3] As example, refer to 21 CFR 178 for use of antistatic and antifogging agents in food-packaging materials.

intended product. Solid compounds are expected to give the lowest risk of interaction with the packaging, but are still at risk.

Examples of possible compatibility-related interactions include:

- Migration of packaging additives/components into the solid material, causing contamination.
- Product degradation induced by adsorption or leaching of impurities.
- Discoloration of the packaging component caused by packaging-related interaction.
- Increase in brittleness/loss of integrity of the packaging component, for example based upon residual solvent incompatibility with the packaging material.

Several case studies which illustrate potential impact to product safety or product elegance due to packaging-related interactions are described below.

2. Overview/Results

2.1. Case I: Contamination by a Migratory Packaging Component

In the first example, a drug substance was packaged into an antistatic primary contact liner. To achieve the desired antistatic properties, a liner that contained an alkyl amide compound was added in a manner compliant with 21 CFR 178. At low levels, this additive is generally regarded as safe for use in food-related packaging (Fig. 18-1). Such additives afford excellent static-dissipating potential to the liners; however, the additives are not covalently bonded to the liner. The additives can migrate throughout the liner to provide the desirable static-dissipating capability. Since the additives are migratory, potential exists for the materials to migrate to the liner surface and subsequently into materials that contact the liner surface. Detection of such components is often enhanced by a high drug product/liner surface area ratio in stability bags. Figure 18-2 illustrates this example, as an unknown impurity peak was observed by HPLC at low wavelength during the execution of a stability study for the drug substance. HPLC-MS analysis of the unknown peak revealed that the impurity was *N*,*N*-Bis(2-hydroxyethyl dodecanamide), an antistatic agent present in the primary liner (Fig. 18-3). Spiking studies with this material further confirmed identity of this material and the source of this impurity as the primary contact liner.

2.2. Case II: Reactivity with a Migratory Packaging Component

In a second example, discoloration of orange-red drug substance powder was observed upon stability (Fig. 18-4). The orange-red colored powder was a hydrochloride salt of an amine-containing compound, and the yellow-orange

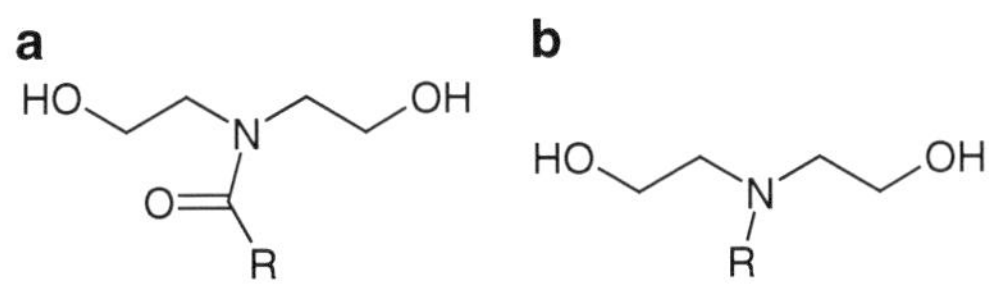

Fig. 18-1. (**a**) *N*,*N*-Bis(2-hydroxyethyl)dodecanamide ($R = C_{12}$) and (**b**) *N*,*N*-Bis (2-hydroxyethyl)alkyl ($R = C_{12}–C_{18}$)amines and are allowed, at low levels, as antistatic and antifogging agents in food-grade packaging materials (21 CFR 178.3130)

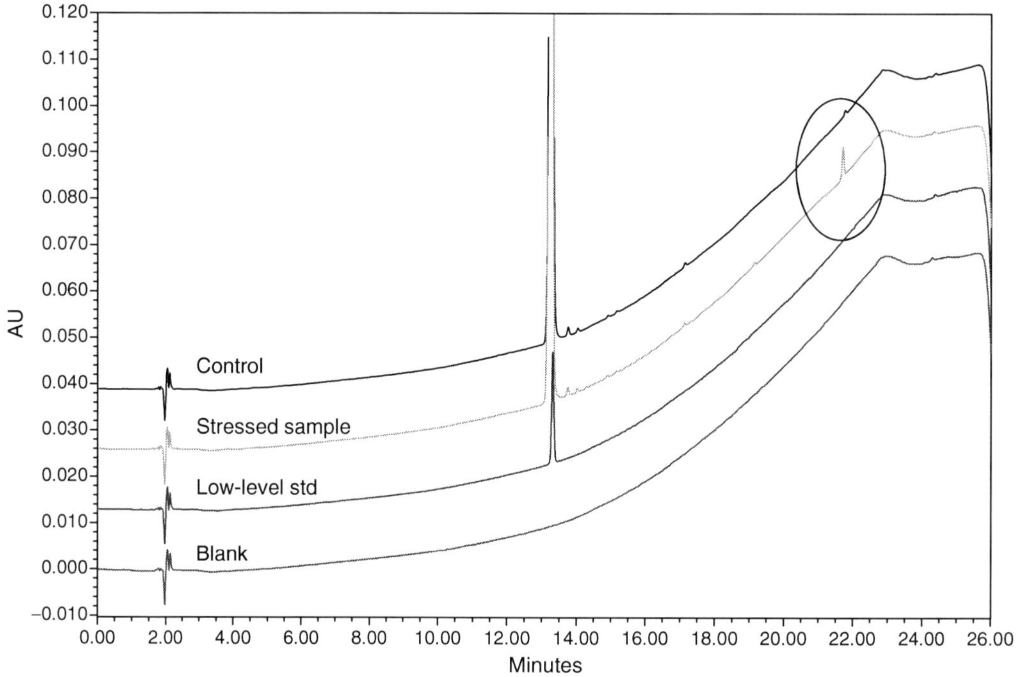

Fig. 18-2. Gradient HPLC chromatograms at 220 nm of four samples (from *bottom* to *top*): blank, low-level standard, 3 month stability sample at 40°C/75%RH, and a control sample. The unknown impurity that was observed is highlighted in the circled region and had a response that was ~0.2% of the drug substance

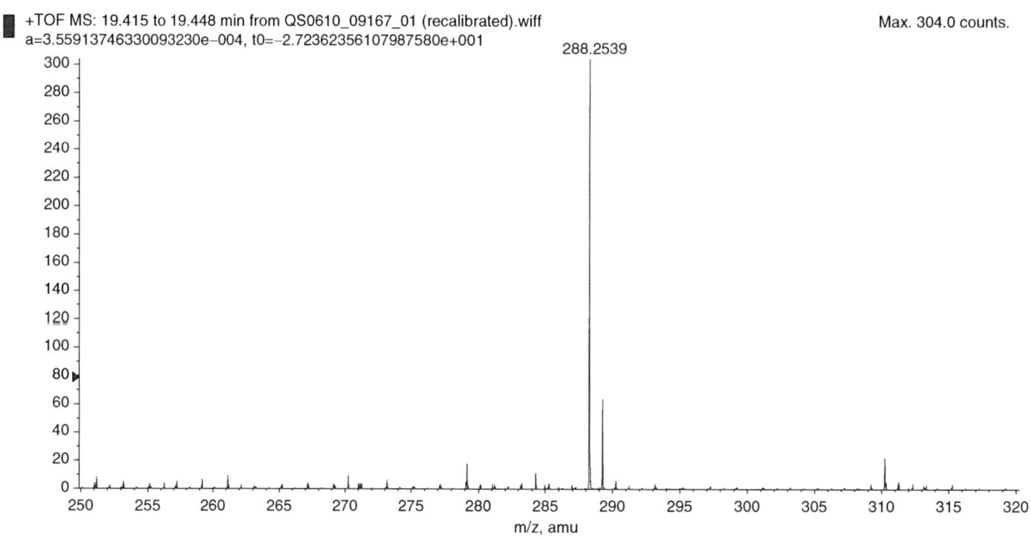

Fig. 18-3. ESI-MS of the unknown impurity. The spectrum corresponded to *N,N*-Bis(2-hydroxyethyl dodecanamide), an antistatic agent in the primary liner

discoloration was observed in material that was packaged in antistatic liners. The yellowish-orange colored material was consistent with the free amine form (free base) of the material (Fig. 18-5) based upon visual observation. Studies indicated that the migratory antistatic component (this time an alkyl

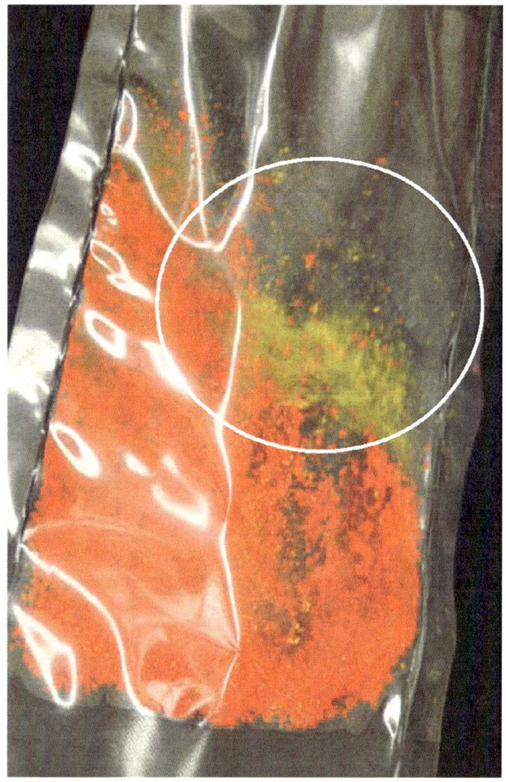

Fig. 18-4. Image of yellowish-orange discolored material observed in a stability bag that initially contained only the orange-red colored material

Fig. 18-5. Yellowish orange appearance of free amine material when compared to the reddish-orange appearance of the amine as the hydrochloride salt

amine additive) interacted with the orange-red hydrochloride salt to form the yellowish-orange free base material. The antistatic amine was able to compete for the hydrochloride and effectively convert a small amount of the drug hydrochloride salt to the free amine form, as observed using FTIR analysis (Fig. 18-6). While the level of the yellowish-orange free amine form would have been very small (ppm level) in production packaging based upon the low antistatic additive level in the liner relative to the amount of packaged product, surface discoloration of the powder at the liner/material interface was easily observed. The clearly evident discolorations could raise quality and elegance concerns about the packaged material regardless of the overall level.

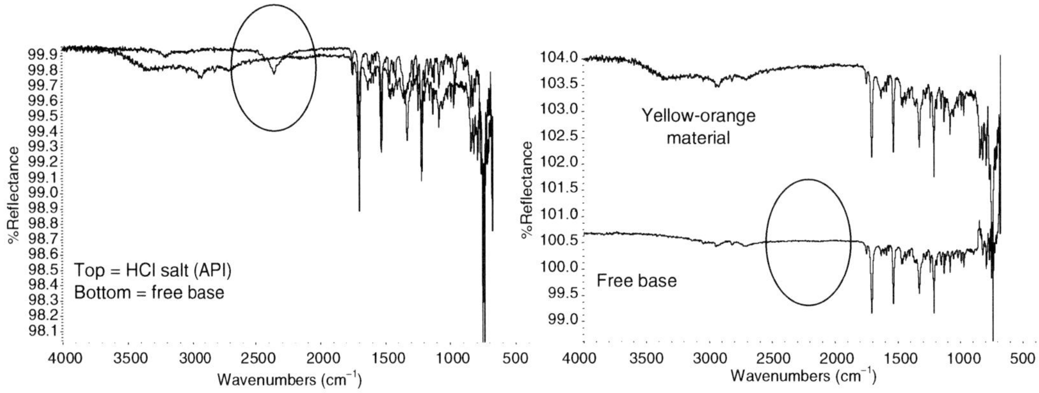

Fig. 18-6. FTIR spectra differentiate between the free amine and hydrochloride salt forms of a material with the absence or presence of the hydrogen-bonded amine (NH^+) stretch at approximately 2,400 cm^{-1}

Fig. 18-7. Brown discolorations which appeared in the creases of the outer layers of the laminated foil liner

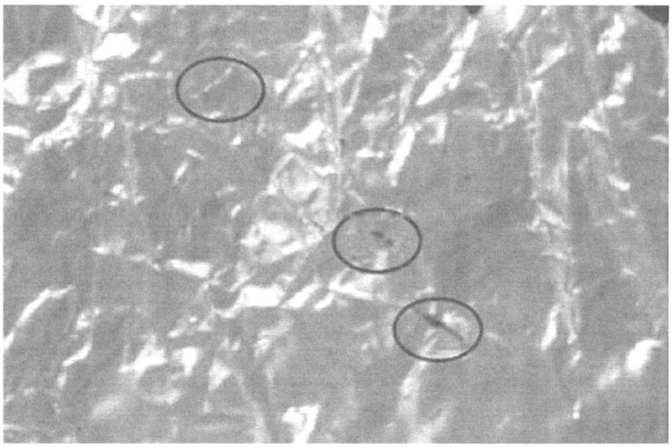

2.3. Case III: Reactivity of the Drug with Packaging Material

In a third example, a low-melting free amine solid intermediate was packaged in the antistatic liner and placed within a laminated foil liner for additional protection during the early stages of compound development. However, during stability studies, it was noted that brown spots began to appear on the outside of the outer laminated foil liner (Fig. 18-7). Meanwhile, the inside liner and product material remained visually unchanged. More careful evaluation of the discolored laminated foil liners revealed tiny pinholes in the creases of the crinkled foil film liner (Fig. 18-8). Furthermore, the discolored spots were located in the same areas as these pinholes, and the spots were limited to the outermost portion of the laminated foil liner (outside the foil layer). The laminated foil liner is comprised of several layers of material, including plastic layers on the inside and outside of the foil film for added strength and puncture/tear resistance.

Studies were conducted in which the outer laminate was exposed directly to the free amine intermediate. The outer laminate quickly became discolored in the presence of the free amine upon heating, indicating incompatibility

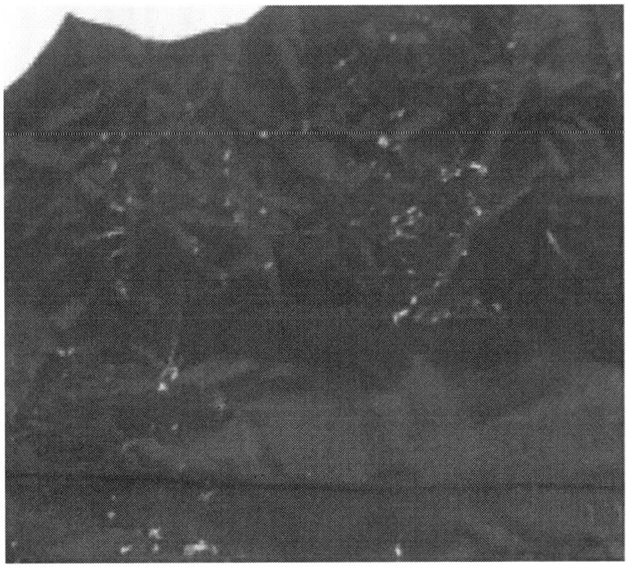

Fig. 18-8. Evidence of pinholes within the laminated foil liner

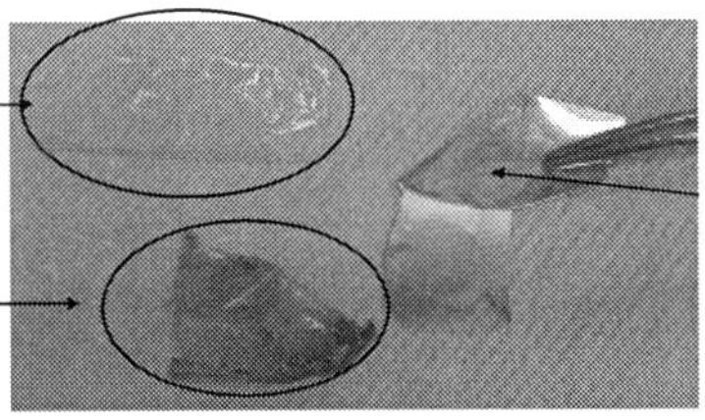

Fig.18-9. Exposure of the outer plastic layer as control (*top*) or discolored when in the presence of material, temperature and moisture (*bottom*)

(reactivity) of the amine with the laminate. This result suggests that the discoloration observed during the stability study was the result of migration of vapors of the free amine through the primary liner and pinholes in the foil film enabling reaction with the outer laminate (Fig. 18-9). In this instance, the discoloration was not observed for the material inside the package nor was the discoloration located next to the product. Nevertheless, elegance concerns would be raised for discolorations that could appear on the outer packaging of a drug material. In this instance, the laminated foil liner was removed as studies showed that the material did not require the additional packaging protection. Instead, a non-permeable drum was utilized along with lower temperature storage to minimize and contain potential product vapors.

2.4. Case IV: Reactivity of the Drug Based Upon Packaging Defects or Packaging-Related Impurities

A number of interactions/reactions with formaldehyde have been presented in the literature. In addition, several cases of formaldehyde reactions have been observed and investigated by the authors (unpublished results). One literature example involves a BMS drug that reacted with formaldehyde which the authors suggest originated from a rubber stopper in a lyophilized formulation

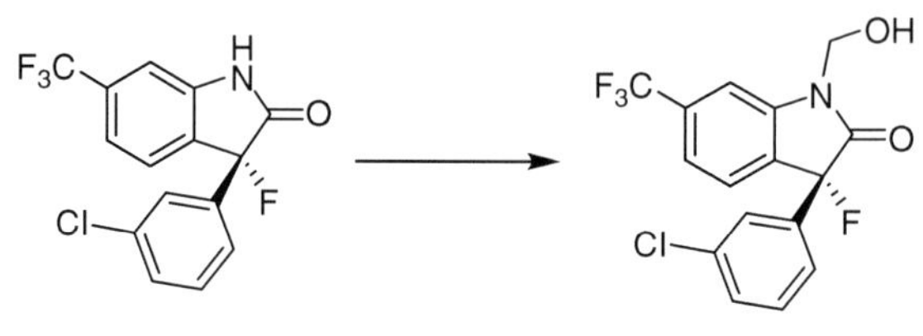

Fig. 18-10. Reactivity of BMS-204352 with formaldehyde from a rubber stopper

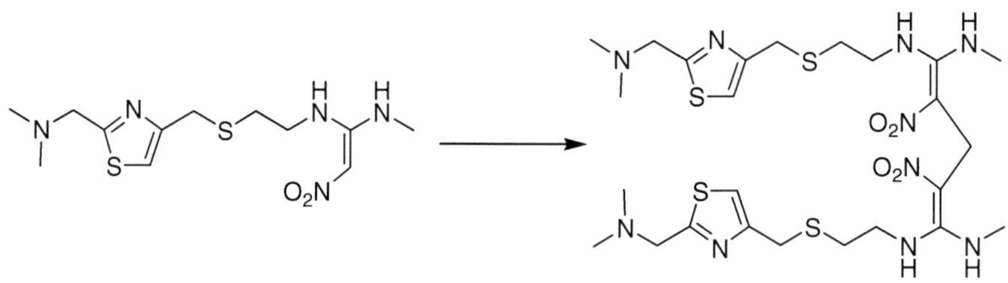

Fig. 18-11. Formation of a methylene-linked dimer of nizatidine in photostressed tablets

(Fig. 18-10) (Nassar et al. 2005). Another literature example that possibly involved formaldehyde is the detection of a famotidine impurity that contained one additional carbon. The impurity was formed only in a child resistant foil pouch laminate. The impurity was synthesized using formaldehyde, however, the authors cautioned that they did not have any direct evidence that formaldehyde was responsible for the impurity (Qin et al. 1994). In a related unpublished example, a methylene-linked nizatidine dimer was observed to form in photostresssed drug product samples (Fig. 18-11). A reaction with formaldehyde was again suspected in this case, but could not be proven. In another unpublished example likely involving formaldehyde, stability samples of white drug substance stored in LDPE liners with laminated foil overwrap exhibited blue spots/lines at the drug-liner interface following storage at 40°C/75% RH for 4 months or 25°C/60% RH for 20 months (Fig. 18-12). Analysis of the blue spots indicated the presence of an impurity known to result from reaction of the drug substance with formaldehyde. It is suspected that formaldehyde from an external source permeated through "pinholes" in the foil overwrap (analogous to the pinholes described in Case III above) to yield the blue discoloration spots.

2.5. Further Examples

At least one observation has been made in which an LDPE-based antistatic liner was discolored after drug substance migrated into the liner during long-term storage.[4] Dissolution of liners with toluene followed by liquid chromatographic (LC) analysis confirmed that the drug had migrated into liners and underwent partial degradation.

[4] Jansen PJ et al. Unpublished results. Eli Lilly and Company.

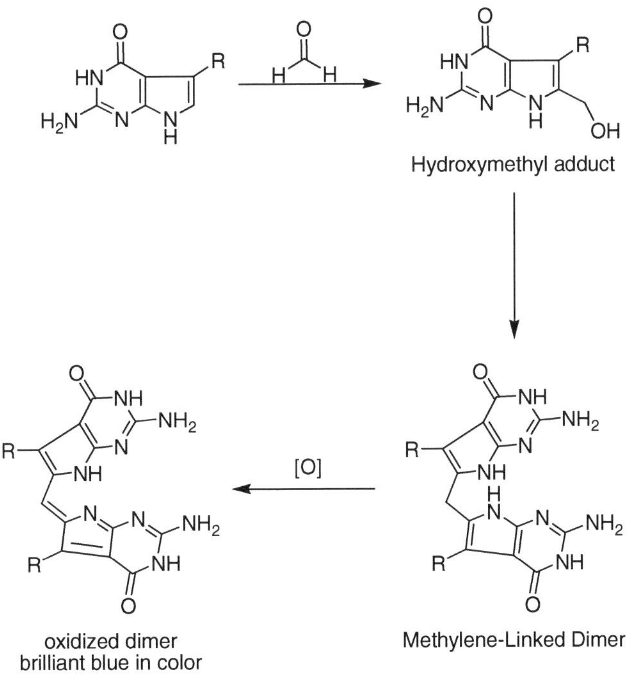

Fig. 18-12. Reactivity of a drug with formaldehyde to form a blue-colored oxidized dimer

In another example, di-*n*-butyl phthalate contamination was observed in tablets stored in LDPE bags during a formal stability study (see Footnote 4). The source of the di-*n*-butyl phthalate was determined to be the secondary cardboard packaging. The di-*n*-butyl phthalate migrated from the cardboard packaging through the primary LDPE liner to the tablet surface.

Finally, bulk tablets shipped in polyethylene bags/polyethylene drums were contaminated with polymer cross-linking agents originating from a foam disk used as a cushion (Maus et al. 2007). It was proposed that acetophenone and 2-phenyl-3-propanol got originated from the polymerization process and diffused from the foam disk through the polyethylene bag to contaminate the product. The authors also proposed that the foam disks used may have been "newer" than the normal and therefore, contained higher levels of the contaminants and that the contaminants originate from the polymerization process used to make the foam disks.

3. Summary

Packaging-related interactions, while infrequent, must still remain a focus of concern for developing thorough product understanding and control for product safety and elegance. LDPE-based primary contact liners in use for small molecule drug substances are effective liners, but are not necessarily inert. Interactions can occur with the primary contact liner, and interactions with the secondary liner or other components can also occur and affect product purity and/or product elegance. Stability studies remain as one critical tool for verifying compatibility between the packaged compounds or products and their corresponding packaging materials as stress stability studies preview what may occur with the bulk packaging stored under more typical conditions. Fortunately, small-scale stability simulators are generally designed to be the worst case simulators (higher liner surface contact area to drug ratio). This high ratio can result in the magnification of potential problems and is good for mechanistic understanding when packaging-related interactions are observed.

References

Balough M, Lennon JD, Hendricker AD, Feinberg TN (2007) LC-GC 25(8):710–717

Fliszar KA, Walker D, Allain L (2006) PDA J Pharm Sci Tech 60(6):337–342

Jaminet F (1968) Inst Pharm Prod Probl Pharm 23(7):375–391

Jenke DR (2002) PDA J Pharm Sci Tech 56(6):332–371

Maus RG, Li M, Clement CM, Kinzer JA (2007) Gas phase transfer of polymer cross-linking agents and by-products to solid oral pharmaceuticals. J Pharm Biomed Anal 45(3):400–406

Nassar MN, Nesarikar VV, Lozano R (2005) Degradation of a lyophilized formulation of BMS-204352: identification of degradants and role of elastomeric closures. Pharm Dev Technol 10(2):227–232

Pellerin F et al. (1991) Interactions and migrations among plastic materials and drugs. Analytical approach. Matieres Plast Usage Pharm, 2nd edn. Ed. Med. Int., Cachan, pp. 273–301

Qin X-Z, Ip DP, Chang KH-C, Dradransky PM, Brooks MA, Sakuma T (1994) Pharmaceutical application of LC-MS. 1 – characterization of a famotidine degradate in a package screening study by LC-APCI MS. J Pharm Biomed Anal 12(2): 221–233

Takacsi N (1989) Orsz Gyogysz Intez 26(5):157–159

Author Biographies

Dr. Mark D. Argentine is a Senior Research Advisor in the Analytical Sciences Research and Development division of Lilly Research Laboratories, Eli Lilly and Company. He received a B.S. in chemistry from the College of William and Mary in Virginia and a Ph.D. in analytical chemistry from the University of Massachusetts, Amherst. He joined Eli Lilly and Company in 1993 and has been involved in analytical control strategy and method development related to synthetic and semi-synthetic drug substance materials. Current responsibilities and interests continue to include the development of analytical control strategies for pharmaceutical analysis, as well as the regulatory aspects of drug development, and quality control.

Mr. Patrick J. Jansen is a Senior Research Scientist in the Analytical Sciences Research and Development division of Lilly Research Laboratories, Eli Lilly and Company. Mr. Jansen obtained a BS degree in Chemistry from the University of Minnesota in 1989 and joined Lilly in 1989. His current responsibilities include leading a team responsible for the design and implementation of ICH stress testing studies for drug substances and drug products, isolation and characterization of degradation products and impurities, and preparation of regulatory documents describing the degradation chemistry of molecules in late-phase development.

Chapter 19

An Overview of Physical Stability of Pharmaceuticals

Yushen Guo

Abstract From drug discovery and development to regulatory approval and marketing, the acceptable stability of the drug substance and drug product is one of the basic quality requirements. The overall stability of a drug product is related not only to the intrinsic chemical stability of the drug molecule, but also the physical form, manufacturing process, interactions among formulation components, package, and storage condition. In this paper, the major physical stability attributes of pharmaceuticals are discussed.

1. Introduction

Stability, as one of the fundamental pharmaceutical quality attributes, needs to be evaluated during the drug discovery and development process, so it can be controlled and maintained for regulatory approval and marketing (ICH Q1A 2003). Change of physical characteristics (e.g., dissolution rate, content uniformity, appearance, taste, and odor) of a drug product, with or without significant chemical degradation, may influence the effectiveness and safety of the medicine, as well as the patient's perception. Stability should be designed into drug products through mechanistic understanding of formulation ingredients, manufacturing processes and container-closure systems (Guo 2008). In the following sections, the physical stability of pharmaceuticals at different stages of drug product life cycle is highlighted (Fig. 19-1).

2. Drug Substance

The understanding of the physicochemical characteristics of drug substance (also known as active pharmaceutical ingredient or API) is the prerequisite for formulation development and design of drug products with optimal stability. Drug substances can experience physical and/or chemical changes under the influences of external environmental factors (e.g., temperature, humidity, pressure, and light). Physical instability of drug substances can be generally classified into two categories based on the main cause (Fig. 19-2). Physical

From: *Pharmaceutical Stability Testing to Support Global Markets*: Pharmasp,
Edited by: K. Huynh-Ba, DOI 10.1007/978-1-4419-0889-6_19,

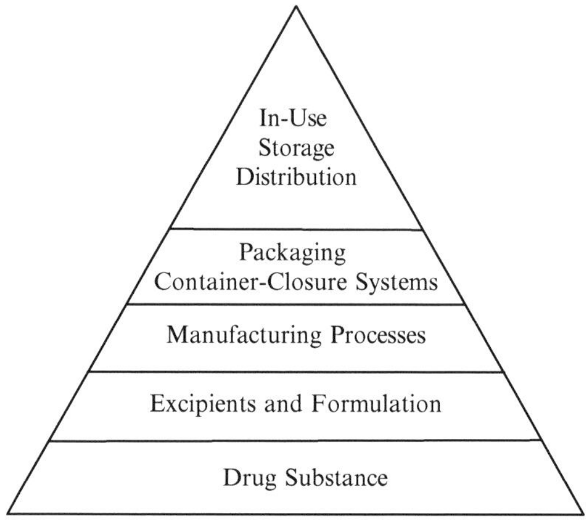

Fig. 19-1. Physical stability pyramid in the life cycle of pharmaceuticals

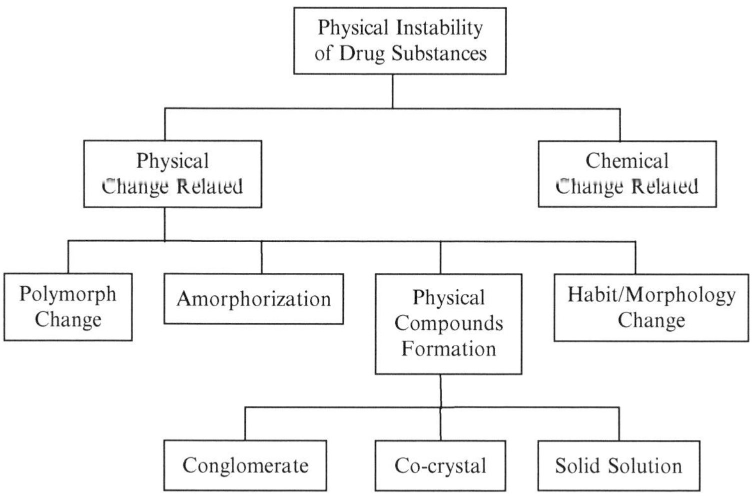

Fig. 19-2. Physical instability of drug substances

changes of drug substance may be induced intentionally or unintentionally during manufacturing processes. The crystal habit particle size, and morphology may be purposely optimized during the process development to obtain the drug substance with desired physical properties. For example, modification of surface habits can affect the physical stability of the suspension formulation in terms of sedimentation and redispersibility. The physical stability and potency uniformity of many topical drug products (e.g., emulsions, creams and ointments) can be affected by API solubility and solid-state properties. Increasing particle size during drug product storage may affect the drug bioavailability or cause physiological irritation.

When two crystalline solids are mixed under mechanical stress or at elevated temperature, conglomerate, co-crystal or solid solution may be formed. Intentional preparation of co-crystal forms of some poorly soluble drug substances has been of increasing pharmaceutical interest in recent years. The differential scanning calorimetric (DSC) and thermomicroscopic methods are often used to study these interactions with minimal material requirement (Davis et al. 2004). On the other hand, the formation of eutectic mixtures between an API and an excipient or between two APIs (e.g., in combination products), which leads to a melting point lower than that of the two components individually, has many implications in dosage form design. Although it has been used occasionally to produce a liquid or low melting point solid to enhance physiologic absorption by increasing the dissolution rate, such as in EMLA cream (eutectic mixture of the local anesthetics lidocaine and prilocaine), eutectic formation may be a concern during high energy formulation processes of most solid dosage forms. Eutectic formation can occur below the eutectic temperature due to pressure-induced increases in the inter-particulate contact between the eutectic-forming components. The impact of certain processing conditions (i.e., roller compaction, tableting) should be evaluated to anticipate any undesired impact on both physical and chemical stability of drug products if eutectic formation is suspected.

The most commonly observed solid state physical changes of drug substances include polymorph conversion, hydration-dehydration interchange, and amorphorization. Phase transition of the API is one of the major concerns during the development and manufacture of solid dosage forms (Morris et al. 2001). The thermodynamically metastable amorphous form may unintensionally be produced during certain pharmaceutical manufacturing processes or observed as the intermediate state for some polymorph changes. Milling or micronization operation of a drug substance, often used for drug product dissolution enhancement or content uniformity purpose, may result in polymorphic form conversion or amorphorization. Solid-state polymorph transformations of drug substances in dosage forms can lead to a change in drug release profiles and other physical characteristics. Thus, it is important to understand the solid-state knowledge space of a new drug substance for development. This includes the mechanisms and kinetics of phase transformations and factors that may influence them. Figure 19-3 shows a general flow chart for polymorph screening and evaluation which should be customized based on the specific drug substance, the targeted dosage form, development stage, and available resources. For amorphous drug substance developed specifically for dissolution enhancement, the study should include not only the drug substance itself, but also the intended formulation and processing methods, such as in multi-component solid-dispersion systems prepared by spray drying or melt extrusion.

3. Excipients and Formulation

Excipients are important functional components of different dosage forms. Not only can they affect the physical transformation of the API, but the excipients themselves may also have different solid-state forms and experience physical changes during the manufacturing processes or storage of drug products. Some commonly used excipients, such as lactose and mannitol, have different polymorphs which have different mechanical and processing properties. The material properties of excipients should be evaluated for their impact on pharmaceutical formulations (Hlinak et al. 2006). Inappropriate use in a specific dosage form can affect the performance of the drug products. Cocoa butter, as a commonly used base for rectal suppositories, is a classic example

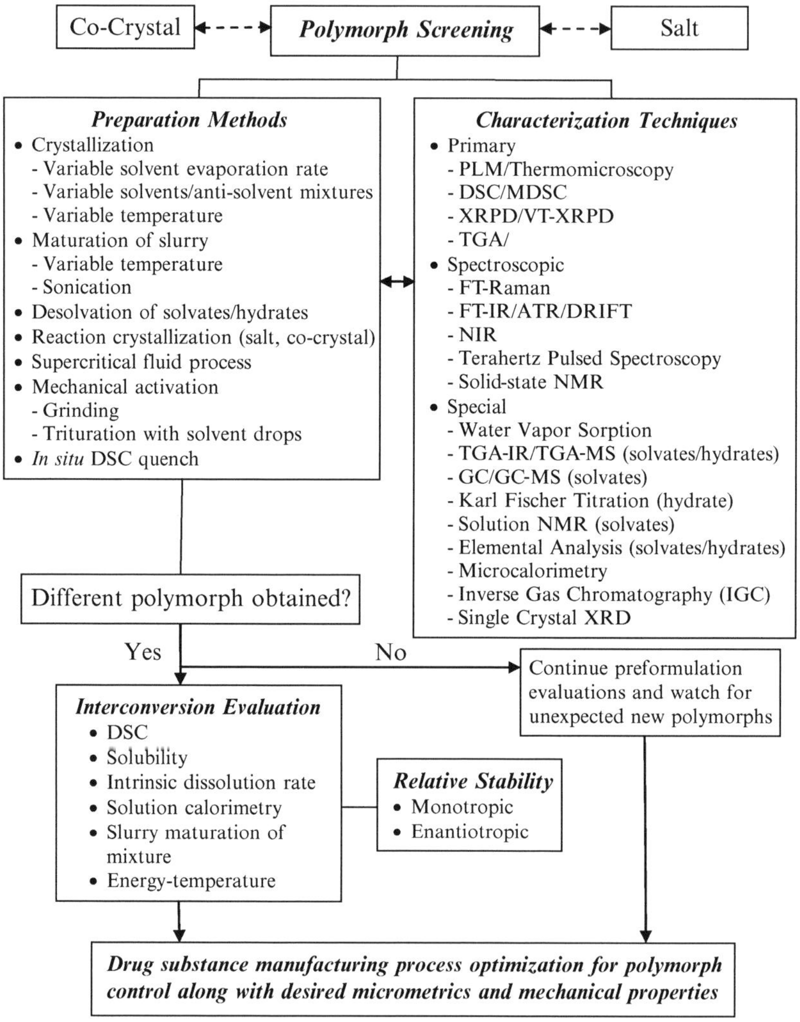

Fig.19-3. Polymorph screening and solid-state physical stability evaluation of API

of the polymorphism effect on the drug product performance. Sometimes the deterioration of excipients and the loss of their functionality under stability storage condition, such as in some special drug delivery systems, can have an adverse effect on the drug quality and safety.

Physical stability of drug products is mostly dosage form specific and requires understanding and controlling of drug substance, excipient and formulation manufacturing processes (Guo 2008). During formulation development, stressed or accelerated stability studies are often used to rank the chemical stability of different prototype formulations in a short time period. It is well known that this approach is less predictive with regard to long-term physical stability, such as suppositories and gelatin capsule formulations stored at 40°C/75% RH. On the other hand, when a good correlation between the physical analysis and conventional stability tests is established, an accelerated physical stability test may be very useful. For example, tests based on the rheology of a semisolid system using heating/cooling cycles (e.g., 40°C/4°C) provided useful information for formulation optimization and selection in a short period of time (Almeida and Bahia 2006).

4. Manufacturing Processes

Physical instability due to phase conversion of drug substances during different pharmaceutical manufacturing processes (e.g., wet granulation and drying) is of continuing industrial and regulatory interest because of potential impacts on the pharmaceutical and biopharmaceutical properties of the final dosage form (e.g., dissolution rate, bioavailability) (Morris et al. 2001). For example, a polymorphic change can happen during granulation and tableting process of the tablet dosage form. During the wet granulation process, one may encounter a variety of interconversions among different anhydrate and hydrate forms of the drug substance. Effective manufacturing processes, including proper equipments and processing parameters, are required to ensure consistent product quality and performance. Process analytical technologies (PAT) can be utilized to monitor the process-induced phase transitions and to optimize manufacturing conditions. For certain drug substances, the solid-state transformation is inevitable due to their intrinsic physicochemical properties. As long as the phase conversions occur consistently and the drug quality can be controlled in a predictable way, it will not affect the development of the drug product.

Temperature is an important parameter in the manufacturing process of many dosage forms. Heat may be generated by interactions among formulation ingredients or by the action of high energy processing equipments. Excessive heat may cause physical and/or chemical changes of the active drug substance and functional ingredients (e.g., preservatives, antioxidants). For example, temperatures fluctuation may cause insoluble ingredients to dissolve, reprecipitate, and change particle size or crystalline form during wet granulation or manufacture of suspension products. In these cases, it is important to control the temperature, not only to facilitate those operations, but also to assure that product stability is not adversely affected.

5. Container-Closure Systems

Medicinal products need to be properly packaged to protect the product from moisture, oxygen, light, and microbiological contamination. When certain drug substances (i.e., carbamazepine) are exposed to excessive moisture, conversion to a less soluble hydrate form can significantly lower the in vitro dissolution rate and oral bioavailability in solid dosage forms. The package can provide a micro-environment to minimize drug instability during distribution and storage. As more and more biological products and new drug delivery systems are being developed, special drug packages are required to ensure long term drug stability. For example, many of the orally dissolving rapid-release drug products are hygroscopic by nature. Delicate oral dissolvable thin filmstrip products, for example, require special package protection to maintain a proper window of relative humidity (RH), since they stick together at high moisture level or high temperature. On the other hand, they become brittle and tend to crack if they are over dried. Aggressive moisture management can cause serious problems in other dosage forms too. For example, the aqueous base coating on the surface of a tablet may begin to crack. It is also well-known that over-desiccation of gelatin capsules can cause esthetic and physical damage.

Package selection of a solid dosage drug product is often based on the physical and chemical stability requirement of the drug product, as well as economic consideration. A successful package provides the optimum balance of cost and performance. A typical conservative approach for launching a new drug with potential stability issues in solid dosage forms is to adapt an overprotective package, such as foil, although this approach is more expensive and often unnecessary. The other approach is to test multiple packaging options simultaneously in the hope of finding at least one lower-cost option that also provides a sufficient barrier. With better understanding of the physical and chemical properties of the API, formulation, and the container-closure options, manufacturers can minimize the number of package designs to be tested while achieving cost-effective protection and acceptable stability.

For liquid dosage form, the interaction of drug substance and excipients with the primary container should be evaluated to ensure no interaction, including physical absorption and extraction of leachables. It has been reported that the interaction of the drug substance with an ammonium sulfate-treated glass vial caused precipitation due to the formation of the less soluble sulfate salt of the drug (Tong et al. 1996). In another case, a minodronic acid injectable solution showed increased particulate matter when the formulation was stored at 25°C, while no potency loss nor particulate increase was observed when stored at 40°C and 60°C (Nakamura et al. 2002). It was discovered that the particulate matter was a complex of minodronic acid and aluminum ions leached from the glass of regular ampoules. The complex formation was exothermic and accelerated stability testing did not predict the long term physical stability in terms of particulate generation.

6. Distribution, Storage and In-use Physical Stability

For sensitive drug products, special handling, storing, and distribution management are required to ensure that the quality and efficacy of the medicine will not be compromised. This should be based on extensive formulation and package development studies to achieve stability optimization of the drug product. The distribution environment of drug products, such as transportation means and seasonal changes, can vary greatly, especially when they are shipped between various climatic zones. Properly designed short-term stressed stability studies could be used to justify the short-term temperature excursions outside the labeled storage conditions during shipping or storage.

Storage conditions are generally less controlled after a medicine has been removed from manufacturer's original container in the pharmacy or stored in the household medical cabinet of patients. In-use stability is especially important for multi-dose products (e.g., oral, topical, and ophthalmic dosage forms) to ensure quality within an acceptable specification once the container is open. Due to repeated opening and closing, there is a risk to its content with regard to microbiological contamination, physical change, or chemical degradation. Stability studies simulating actual in-use conditions are essential to establish the in-use shelf life and storage condition recommendations. New moisture and oxygen control technologies can help to maintain drug stability during use. For example, sorbent can be placed under the cap of the drug bottle to maintain the dry atmosphere inside. Also, resin layers containing moisture absorbers can be co-extruded into bottles which provide better moisture protection.

Co-administration of i.v. admixtures of two or more drugs using Y-site injection is very common in hospital settings. Therefore, both chemical and physical incompatibility (e.g., discoloration, precipitation) should be evaluated. In the case of physical incompatibility, the drugs may be administered separately, or the i.v. tubing should be flushed thoroughly before administering the second drug.

7. Interplay Between Chemical and Physical Stability

When investigating physical instability of pharmaceutical systems, it is important to understand the root causes so that effective actions can be taken to resolve the problem. Chemical degradations are often found to play an important role in this aspect and may happen during different stages of the drug life cycle. Many physical changes, such as organoleptic properties, are caused by chemical change of the drug substance or excipients in the formulation during storage. For example, oxidative degradation reactions or Maillard reactions between certain active pharmaceutical ingredients and reducing sugars can lead to gradual coloration of the drug products. The vinegar odor of the old aspirin tablet is due to the release of acetic acid from the hydrolysis degradation. Gelatin capsule cross linking under accelerated stability conditions is a well-known phenomenon that results in reduced drug dissolution. Trace formaldehyde from some excipients (e.g., corn starch) are the main cause for the adverse effect on in vitro dissolution rates due to cross links of lysine and arginine residues in gelatin (Digenis et al. 1994). The incorporation of both glycine and citric acid in some formulations has been shown to prevent pellicle formation or cross linking of the gelatin capsule upon storage under accelerated conditions (Adesunloye and Stach 1998). Particulate formation of an antitumor agent solution following agitation was found to be associated with the formation of a poorly soluble aldehyde impurity caused by photo-degradation (Rubino et al. 1999). This could be prevented by protecting the drug solution from light during chemical synthesis and production of the dosage form, formulation pH optimization, and use of protective container and package systems.

As to the chemical degradations initiated by physical changes, the most common examples occur when a crystalline drug substance is unintentionally converted totally or partially to the metastable amorphous form during a manufacturing process, as described in the previous sections. For some liquid dosage forms, a pH shift, which initially may be the result of a chemical change itself, can further accelerate the chemical degradation reactions of both active and/or other functional ingredients.

8. Conclusion

Physical stability of pharmaceuticals is an integral part of drug stability and quality requirements. It can have a potential impact not only on the esthetic appearance, but also on the performance, efficacy, and safety of a drug product. Along with the chemical stability, acceptable physical stability should be designed into the drug product based on the understanding and selection of drug substance forms, excipients, formulations, and manufacturing processes. The physical and chemical stability of a drug product also need to be maintained with suitable container-closure system during distribution and storage until taken by patients.

References

Adesunloye TA, Stach PE (1998) Effect of glycine/citric acid on the dissolution stability of hard gelatin capsules. Drug Dev Ind Pharm 24:493–500

Almeida IF, Bahia MF (2006) Evaluation of the physical stability of two oleogels. Int J Pharm 327:73–77

Davis RE, Lorimer KA, Wilkowski MA, Rivers JH (2004) Studies of phase relationships in cocrystal systems. ACA Trans 39:41–61

Digenis GA, Gold TB, Shah VP (1994) Cross-linking of gelatin capsules and its relevance to their in vitro-in vivo performance. J Pharm Sci 83:915–921

Guo Y (2008) Impact of solid-state characteristics to the physical stability of drug substance and drug product in handbook of stability testing in pharmaceutical development (Kim Huynh-Ba ed.), Springer, New York, NY

Hlinak AJ, Kuriyan K, Morris KR, Reklaitis GV, Basu PK (2006) Understanding critical material properties for solid dosage form design. J Pharm Innovation 1(1):12–17

ICH guidance Q1A (R2) (2003) Stability testing of new drug substances and products http://ww.w.ich.org

Morris KR, Griesser UJ, Eckhardt CJ, Stowell JG (2001) Theoretical approaches to physical transformations of active pharmaceutical ingredients during manufacturing processes. Adv Drug Deliv Rev 48:91–114

Nakamura K, Yokohama S, Sonobe T (2002) Failure of stability prediction for minodronic acid injectable by accelerated stability testing. Int J Pharm 241:65–71

Rubino JT, Chen LL, Walker JT, Segretario J, Everlof JG, Hussain MA (1999) Photoinduced particulate matter in a parenteral formulation for bisnafide, an experimental antitumor agent. Pharm Dev Tech 4:439–447

Tong WQ, Clark J, Franklin ML, Jozwiakowski JP, Lemmo JB, Sisco JM, Whight SR (1996) Identification and characterization of the sulfate precipitate in GI147211C IV formulation. PDA J Pharm Sci and Tech 5:326–329

Author Biography

Dr. Yushen Guo currently is employed at Achaogen, where he is responsible for preformulation and formulation development activities of new drug candidates. He has over 20 years of chemical research and pharmaceutical development experience. His research interest includes solid-state characteristics of drug substances and impact on drug products, API forms and crystallization process development, solubilization of poorly soluble drug compounds, drug stability and degradation mechanisms, and new drug delivery systems. He received his Ph.D. in Physical Organic Chemistry from Iowa State University and then completed 2 years of postdoctoral research on solid-state pharmaceuticals in the Pharmacy School at University of Wisconsin-Madison.

Chapter 20

Stability of Split Tablets

Vilayat A. Sayeed, Abhay Gupta, and Mansoor A. Khan

Abstract The current pharmacy practice of splitting tablets has grown exponentially in the last few years to contain the ever growing cost of health care. At present, there is no scientific evidence in the public domain to demonstrate that the random splitting of tablets would not pose any safety concerns by compromising the quality of the product. The presentation would addresses the quality risk associated with the tablet splitting and the need for generating adequate scientific data before recommending a tablet splitting.

1. Introduction

In the product approval process, drug products are tested under the storage condition specified in the labeling to meet the established quality specification over their intended expiration period. However, once they reach the market and it gets into the hands of the consumer, they can be stored and used on the conditions which are actually not recommended under the label. Tablet splitting is one such practice that is commonly used by the consumers and in some cases by major health care providers. To get a broader understanding on this issue, an internet search on the word "split tablet" was done and this resulted in over a million hits highlighting the interest among consumers, regulators and industry about this subject. In most cases, the concern was on the uneven split of the tablet resulting in an inaccurate dosing. This practice of splitting tablets when not specified in the labeling is not recommended by the Food and Drug Administration, as this can potentially have an adverse effect on the quality of the drug product. However, a number of products are designed to split, in such cases, the tablets have a score-line on one or both sides of the tablet to assist a relatively even tablet splitting. For scored products, this information is part of the product label, and it generally states that the split tablet must be used within a short period of time after splitting. But in practice, the pharmacists may typically split 30 days of tablets in their office prior to dispensing. This practice leads to the storage of a scored split tablets over an extended period of time beyond the intent of the label and splitting un-scored tablets which is

From: *Pharmaceutical Stability Testing to Support Global Markets*: Pharmasp,
Edited by: K. Huynh-Ba, DOI 10.1007/978-1-4419-0889-6_20,

not specified in the labeling. In either scenario, the product is being stored and being used under conditions that are not recommended in its labeling.

The concern becomes even greater if the products are not scored and is being split prior to use. In limited number of cases, the product is split for medical reasons to titrate the correct dose regimen. However, the major reason for tablet splitting is to control prescription drug costs. To control the cost, a number of insurance companies actively promote this practice and are offering free tablet-splitters, no co-pay or other incentives to convince patients to purchase higher strength tablets – which may many times cost about the same as a lower strength of the medicine – and then break the tablets in half or quarter doses. This practice though widely used as a cost containment measure, lacks scientific knowledge in the public domain to support that these products can safely be split, stored under use conditions and used for an extended period of time. Hence it is important to understand the risks associated with this practice.

The most commonly cited problem when the tablets are split is that of weight variation among the split halves (McDevitt et al. 1998; Rosenberg et al. 2002; Teng et al. 2002) Studies on the effect of tablet splitting on in-vitro dissolution profile have yielded varying results. For some sustained- or extended-release drug product, tablets splitting had no effect while for others an increase was observed in the rate of drug release from the split tablet (Skoug et al. 1991; Mandal 1996; Shah et al. 1987; Simons et al. 1982; Erramouspe and Jarvi 1998). Several clinical studies on the effectiveness of statin drugs (atrovastatin, lovastatin, simvastatin) have shown that tablet splitting does not negatively affect the levels of low density lipoproteins, total cholesterol and/or triglycerides in patients (Duncan et al. 2002; Parra et al. 2005) However, none of these studies looked at the impact of this practice on the product quality, i.e., assay, dissolution, impurity profile, etc.

The study was undertaken to understand the impact of this practice on the product quality by looking beyond just weight variation and product dissolution. Gabapentin tablets, 600 mg were selected for this study as these products were initially available as un-scored tablets, but were later marketed as scored tablets without any change in the formulation compositions. Furthermore, quality standards related to the impurities and degradation for this drug product exists in public domain (USP) allowing the information to be shared in the public forum and the drug product is available from multiple suppliers. Additionally, an OTC sustained-release product was also evaluated to study the influence of tablet splitting on dissolution profile of a commonly used OTC product.

Gabapentin [structure I] drug product was approved initially as an anticonvulsant and later on as an analgesic and neuroleptic. It is know to degrade via intramolecular cyclization to form a lactam: 3,3-pentamethylene-4-butyrolactam (2-azaspiro[4, 5]decan-3-one) [lactam, structure II] and the USP limit for this impurity is 0.4% over the shelf life of the drug products.

2. Materials and Method

2.1. Materials

Bulk containers of gabapentin tablets (600 mg) were obtained from three different manufacturers (G1, G2 and G3). The drug products G1 and G3 were scored, while drug product G2 was not scored. The OTC sustained the release product that has been used in this study contained four active ingredients.

2.2. Sample Preparation

Tablets were split into two parts using a tablet-splitter. Gabapentin tablets were dispensed into amber pharmacy vials, which were capped and placed in environmental chambers under long-term storage conditions (25°C/60% RH) or intermediate stability conditions (30°C/60% RH) for 9 weeks. Samples were removed from the environmental chambers at 0, 2, 4, 6 and 9 weeks and analyzed for potency, impurity levels, hardness and dissolution. The OTC sustained release product was split and tested immediately for dissolution.

2.3. Sample Analysis

Twenty intact tablets and 40 split portions were weighed to determine the weight variation of the gabapentin drug products. These tablets were then ground in a mortar with a pestle to a fine, homogenous powder for potency testing and analyzed using validated HPLC method. A Tablet Hardness Tester was used to determine the hardness of 10 intact and 20 split gabapentin tablets. Dissolution testing was performed using the methods outlined in the current USP-NF for both gabapentin drug products and the OTC combination product. The dissolution samples were analyzed using validated HPLC methods.

3. Results

The weight variation of intact tablets was less than 0.8% for all three gabapentin drug products. However, the weight variation was significantly higher for the split tablets, with the split halves of the two scored gabapentin drug products (G1 and G3) showing higher weight variation of 6 and 3% respectively as compared to 2% for the split halves of the un-scored G2 drug product.

No difference was observed between the potency results of the whole and split gabapentin tablets from the three manufacturers (Fig. 20-1). The potency of all drug products stayed within the specifications of 90–110% of the labeled claim throughout the study period under both storage conditions, though a slight decline in potency was observed for the samples stored at the 30°C. The results suggest that the temperature has a much bigger impact on gabapentin drug product quality as compared to tablet splitting. The lactam impurity levels also showed an increase for these samples stored at 30°C, with the G3 tablets (whole and split) failing to meet public standard at 9 weeks (Fig. 20-2).

A significant decrease in the hardness of G1 tablets was also observed when split, with hardness values falling to 10–15 KgF range from >20KgF, suggesting significant influence of tablet splitting on product quality of G1 tablets.

No influence of tablet splitting, and storage condition/time was observed on the dissolution profile of gabapentin drug products and all products meet the public standard for dissolution testing. However, three out of four active ingredients present in the OTC combination sustained-release drug product showed a faster dissolution with split tablets as compared to the whole tablets (Fig. 20-3).

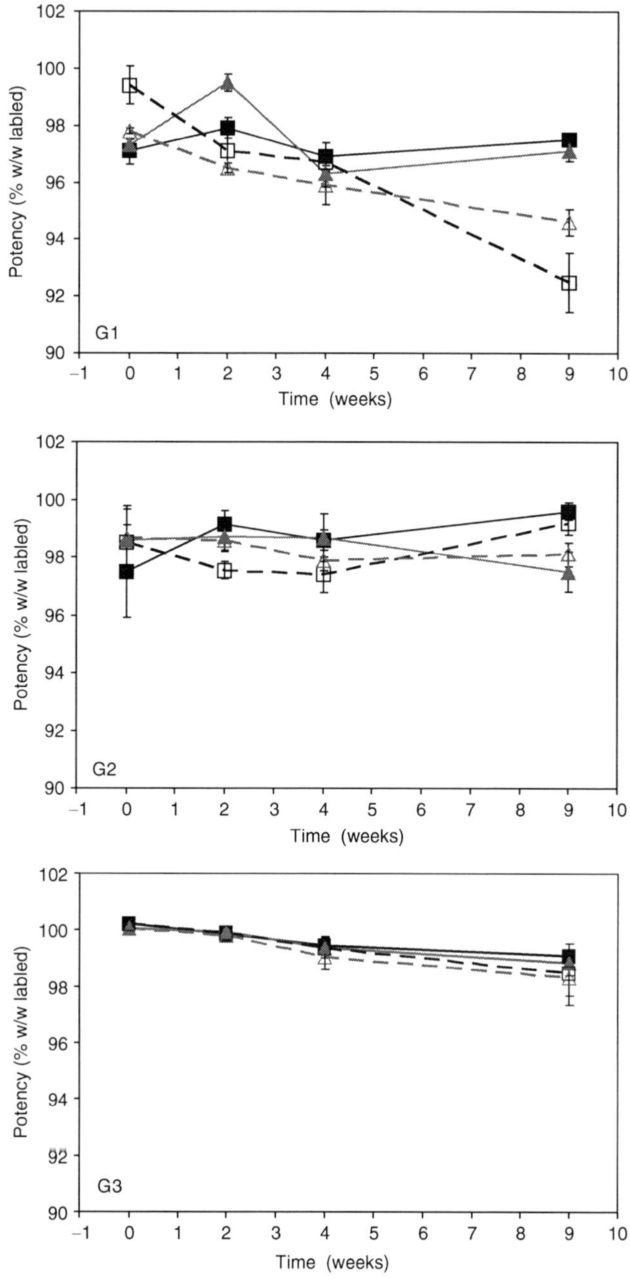

Fig. 20-1. Change in gabapentin potency for the whole (*squares*) and split (*triangles*) gabapentin drug products over 9 weeks under 25°C/60% RH (*solid line*) and 30°C/60% RH (*dashed line*) storage conditions

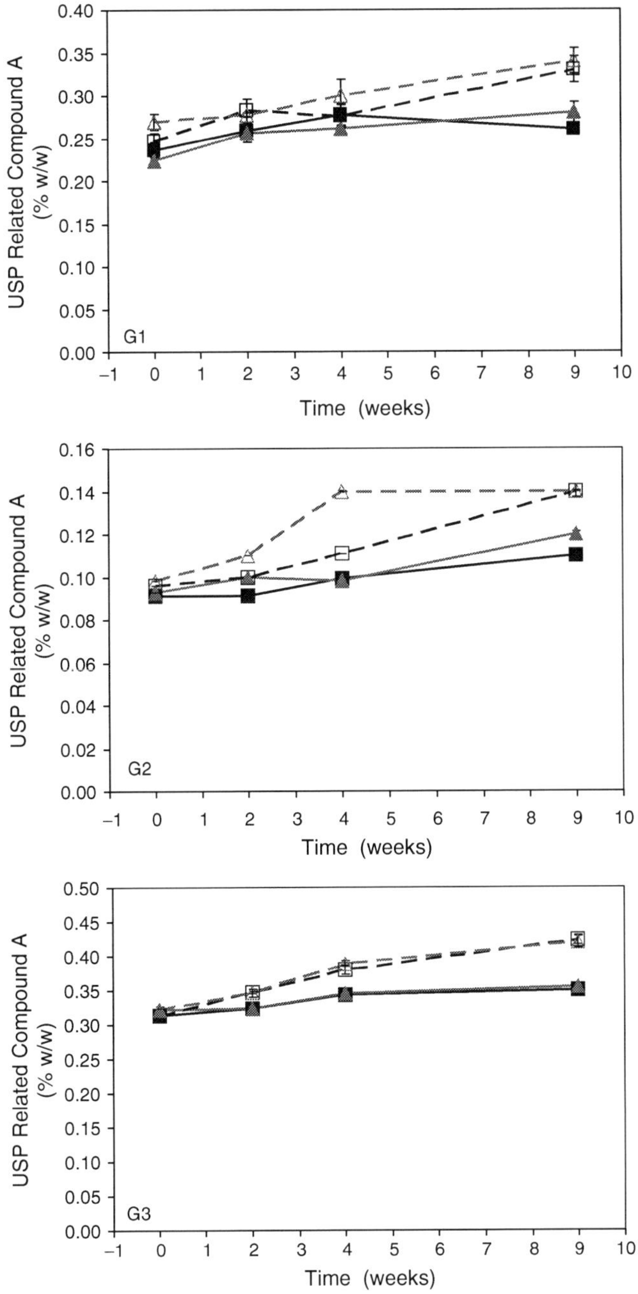

Fig. 20-2. Change in lactam concentration for the whole (*squares*) and split (*triangles*) gabapentin drug products over 9 weeks under 25°C/60% RH (*solid line*) and 30°C/60% RH (*dashed line*) storage conditions

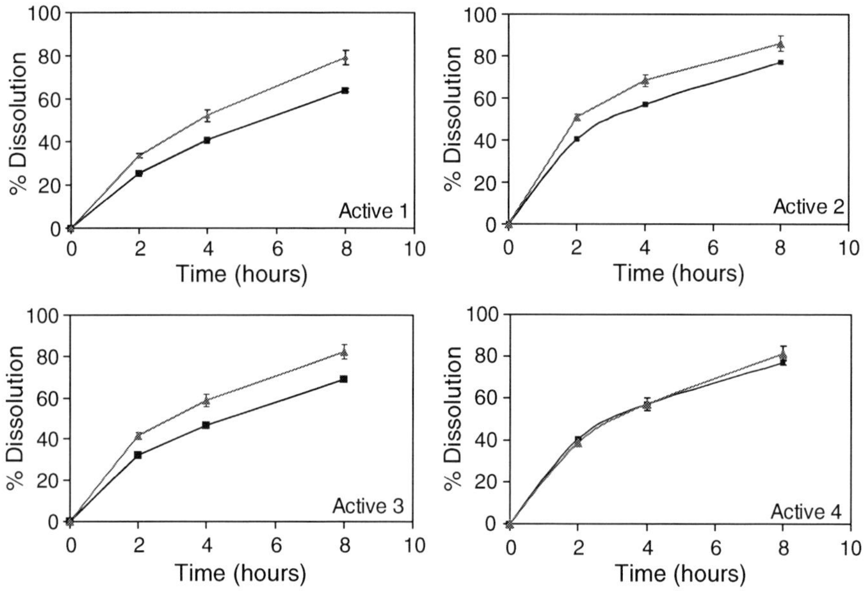

Fig. 20-3. Difference between dissolution profiles of the whole (*squares*) and split (*triangles*) tablets of the OTC combination products containing four active ingredients

4. Conclusions

Splitting of G1 and G2 tablet did not significantly impact the quality of these drug products. The concentration of the gabapentin impurity in these two drug products showed an increase with time, but met the public standard. However, the impurity levels in G3 failed to meet the public standard indicating that other factors, such as formulation and manufacturing process can make the same product from one supplier more susceptible to degradation than from the other supplier. The dissolution of gabapentin drug products studied was not affected by tablet splitting. However, the split OTC combination tablets showed a significant faster dissolution for three out of four active ingredients.

The study shows,[1] tablet splitting (split-tablet weight variability) has a potential for significant variability in dosing regimen, product quality, pharmacokinetic endpoints and clinical outcome. This finding demonstrates that the same product from different manufacturer may behave differently, when split and stored under use conditions and thus is not created equal. Therefore the results of this study should not be extended to other drug products or in no way can be concluded that the quality attributes of split tablets are fully understood.

---

[1]This work was done in the FDA laboratories of Division of Product Quality Research, Office of Testing and Research located in White Oak, Silver Spring and findings were published in Int J Pharm (2007) 350(1–2):65–69 and Am J Health Syst Pharm (2008) 65(24):2326, 2328.

References

Duncan MC, Castle SS, Streetman DS (2002) Effect of tablet splitting on serum cholesterol concentrations. Ann Pharmacother 36:205–209

Erramouspe J, Jarvi EJ (1998) Effect on dissolutionfrom halving methylphenidate extened-release tablets. Ann Pharmacother 32:1372–13723

Mandal TK (1996) Effect of tablet integrity on the dissolution rate of sustained-release preparations. J Clin Pharm Ther 21:155–157

McDevitt JT, Gurst AH, Chen Y (1998) Accuracy of tablet splitting. Pharmacotherapy 18:193–197

Parra D, Beckey NP, Raval HS, Schnacky KR, Calabrese V, Coakley RW, Goodhope RC (2005) Effect of splitting simvastatin tablet for control of low-density lipoprotein cholesterol. Am J Cardiol 95:1481–1483

Rosenberg JM, Nathan JP, Plakogiannis F (2002) Weight variability of pharmacist-dispensed pill tablets. J Am Pharm Assoc 42:200–205

Shah VP, Yamamoto LA, Schuirman D, Elkins J, Skelly JP (1987) Analysis of in vitro dissolution of whole vs. half controlled release theophylline tablets. Pharm Res 4:416–419

Simons KJ, Frith EM, Simons FE (1982) Dissolution and bioavailability studies of whole and halved sustained-release theophylline tablets. J Pharm Sci 71:505–511

Skoug JW, Borin MT, Fleishaker JC, Cooper AM (1991) In vitro and in vivo evaluation of whole and half tablets of sustained-release adinazolam mesylate. Pharm Res 8:1482–1488

Teng J, Song CK, Williams RL, Polli JE (2002) Lack of medication dose uniformity in commonly split tablets. JAMA 42:195–199

Author Biographies

Dr. Abhay Gupta has a BS and MS in Pharmacy from India and obtained his Ph.D. in Industrial and Physical Pharmacy from Purdue University, West Lafayette, IN. Dr. Gupta joined the US Food and Drugs Administration in the Division of Product Quality Research as a pharmacologist in 2004, where he has won numerous awards for his research dealing with pharmaceutical product quality. He has presented his work in national and international meetings and has published over 15 scientific articles in peer-reviewed journals. He has co-authored 2 book chapters, and serves as a reviewer for a number of pharmaceutical journals including *Journal of Pharmaceutical Sciences, Pharmaceutical Research, International Journal of Pharmaceutics* and *Journal of Pharmaceutical and Biomedical Analysis*

Dr. Mansoor A. Khan is the Director of Division of Product Quality Research in Center for Drug Evaluation and Research at US Food and Drug Administration. Prior to joining FDA, Dr. Khan was a Professor of Pharmaceutics and Director of Graduate Program in the School of Pharmacy at Texas Tech University Health Sciences Center. He is a registered pharmacist, and has earned his Ph.D. degree in Industrial Pharmacy from the St. John's University School of Pharmacy at New York in 1992. He has published over 160 peer-reviewed manuscripts, four texts including the "Pharmaceutical and Clinical Calculations," 8 book chapters, and more than 125 poster presentations and over 100 invited oral presentations in various meetings. Dr. Khan graduated ten Ph.Ds in pharmaceutics. He is the Chair-elect of Formulation Design and Delivery section of AAPS, and has been recognized as AAPS Fellow.

He serves on the editorial board of *Pharmaceutical Technology* and the *Journal of Clinical Research and Regulatory Affairs*.

Dr. Vilayat A. Sayeed received his Ph.D. in Organic Chemistry from University of Manitoba, Winnipeg, Canada in 1982, worked as Postdoctoral Research Associate at University of Toronto and University of Manitoba, Canada for 2 years. In 1984, he joined Pharm-Eco Laboratories as a chemist and was later promoted to principle investigator on NCI contract. His work involved research and process development actives to support the Phase II, Phase III clinical trials for the National Cancer Institute. He joined the Office of Generic Drugs, Center for Drug Evaluation and Research in 1992 as review chemist, was promoted to Team Leader in 1996 and in 1999 was promoted to Deputy Director, Division of Chemistry II. He assumed his current position in 2004 as the Director, Division of Chemistry III.

Chapter 21

Temperature Monitoring During Shipment and Storage

Conny Axelsson

Abstract AstraZeneca has rolled out a global Temperature Monitoring Program. The presentation will focus on the reasons for having taken that approach, the various challenges we met ,and also some learning points and recent changes in our program.

AstraZeneca believes that it is a good ethic to keep control of the products during transportation between the AstraZeneca sites; it also gives us information on the transport routes and allows us to continually improve our transports. Even though cool chain transports are challenging, we found room temperature transports even more challenging mostly due to complexity between markets and product submissions. It's key to be pragmatic, keep it simple, have clear roles and responsibilities and have the same information at hand.

1. Introduction

AstraZeneca implemented a temperature monitoring program covering most deliveries and transportation routes for Active Pharmaceutical Ingredients (APIs), bulk formulated product and finished packed product, more than 140 routes and 130 products are included.

This program helps in assuring product quality through its shelf life by evaluating, handling and preventing temperature excursions.

2. Global Monitoring

Global temperature monitoring, not only for cold chain products, turned out to be a good distribution practice since authorities recently started to pay attention into this area. This is leading edge, AstraZeneca has been one of the first companies to go with this, and others just follow or watch with interest. Now, there are sophisticated technologies available allowing us to do the temperature monitoring accurately and easily around the globe.

From: *Pharmaceutical Stability Testing to Support Global Markets*: Pharmasp,
Edited by: K. Huynh-Ba, DOI 10.1007/978-1-4419-0889-6_21,

The goal is to have visibility and control over the shipping conditions throughout the supply chain within AstraZeneca.

When we started, we did not fully understand the complexity. Global products mean different label claims in different markets, and also market specific requirements and regulations. Room temperature could mean 2–25°C in some markets, up to 30°C in some markets and below 30°C in others.

Add to that the variety of products present at these conditions; API's, dry powders, tablets, capsules, ointments and liquids, and also different packages as glass and plastic ampoules, blisters, bottles and bulk containers. The only way to manage the complexity is to keep processes simple and by using risk assessments.

2.1. AstraZeneca's Approach

We used a pragmatic approach and the products were divided into three classes; the 2–8°C, 2–15°C, and 2–25°C. The first two classes are temperature controlled and monitored throughout the shipment process, while the third class is only monitored since climate controlled containers/transports are expensive to be used for all cases and room-temperature products are less sensitive than the cold chain products.

Products can be shipped in a "lower" temperature class than originally assigned. However, it could be critical for some products such as emulsions, liquids and effervescing tablets to be shipped in subzero conditions.

The Mean Kinetic Temperature (MKT) is calculated for all shipments and used for evaluation.

Fig. 21-1. TSS temperature monitor. Temp Tracer is the trademark of Temperature Sensitive Solutions AB, Sweden

The data monitored is stored in a global database available to all sites; this makes the evaluation easy because all sites have access to the same data.

3. Tools and Equipment Involved

3.1. The Tools to be Used

Mainly, there are three things needed in this process; a temperature monitor, a verifier, and a reader plus its software.

An example for the temperature monitor would be that supplied by TSS, Sweden. (see Fig. 21-1). It is the size of a credit card (2″×3″). It is failsafe, so it does not take any false readings, and it can be used with multiple shipments.

Each of these cards has a unique number to identify it, and they are calibrated, and assigned a 1-year lifecycle. Typically, two of these monitors are used with each shipment.

3.2. The Way of Using the Tools

Three steps are involved in the process; create a profile, launch the monitor, stop it and read the data at the end of the shipment. A profile contains three key bits of information; a route detail, sampling intervals and temperature limits. After specifying those, the profile is stored into the global database, a profile only needs to be created once. The profile is loaded onto the monitoring device and initiated. The monitor is then placed on the consignment and reads the temperature all the way to the receiving site, it is then stopped, and the data are downloaded for analysis.

3.3. Data Output

An example of the sort of data extracted from this type of devices can be viewed in Fig. 21-2. It shows two excursions throughout the shipping process; one of them was over the 25°C maximum limit but for brief moments, the other was under zero degrees and it was for a few hours.

The output is very visible and allows us to drill into the data and investigate excursions easy, since we have all data at hand and can track it throughout the shipment identifying, by contacting the freight forwarder, exactly what happened during the shipment.

What happened was, when the product landed in Frankfurt – at the mark number 2 – it was probably subject to some sunlight for a brief moment, which caused that spike in the reading passing 25°C.

As for the second excursion, it turned out that when the shipment arrived to the warehouse – around mark 4 – it was put in a freezer instead of a cold room, causing that below-zero excursion.

4. Evaluation of Temperature Excursions

By definition, an excursion is a period of time during transportation when the shipment/products are exposed to temperatures that are outside the temperature limits set on the monitors. It is important to remember that the limits on the monitors are action limits and further assessment is required. It might be well that we have supporting data and we can release the product.

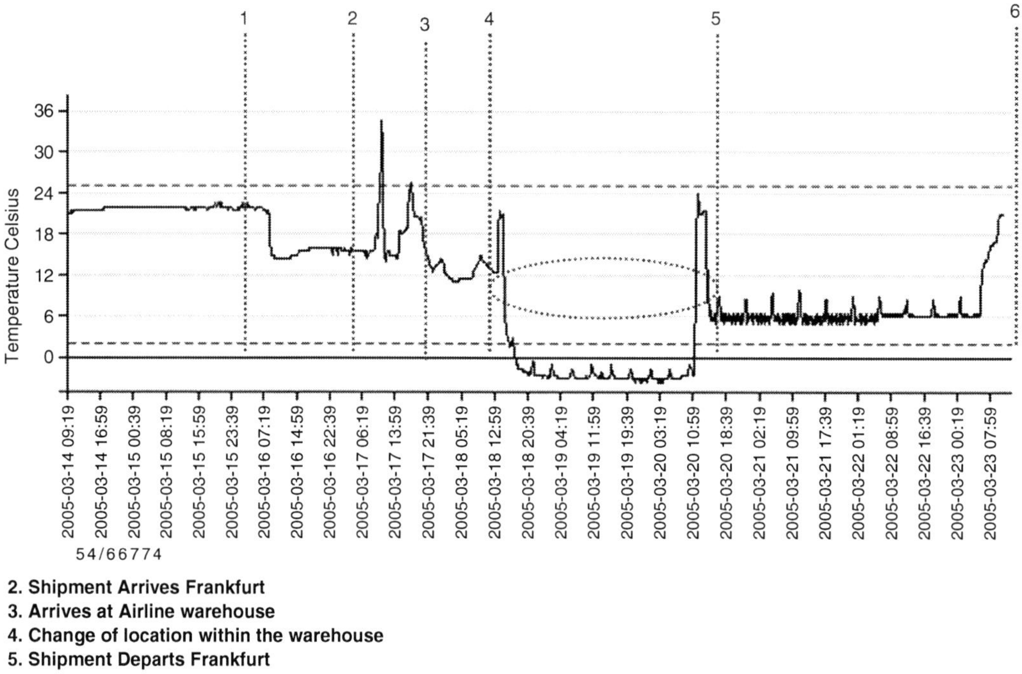

Fig. 21-2. Temperature monitor data output

4.1. Excursion Evaluation Step 1

The receiving site consults the global database and asseses the data. For the product exemplified above, excursion temperatures above 25°C are accepted under certain conditions as long as the MKT is between 2 and 25°C. Excursions below 2°C are not acceptable by any means.

4.2. Excursion Evaluation Step 2

In the second step, when it has been evident that it is a true excursion, the receiving site raises a complaint in the global complaint database to the manufacturing site of this product, which is also responsible for doing an investigation with input from distribution and stability experts if necessary.

All sites are linked up to the same database to see the same data, so that they can start evaluating what has actually happened and give input to the investigation.

In this particular case, the product was rejected due to the sub zero excursion.

5. Prevention of Temperature Excursions

5.1. Quality Systems

It is important to have global quality guidelines giving very clear rules and responsibilities to both the sending site and the receiving sites and the people working to investigate excursions if and when they happen.

5.2. Understanding of the Supply Chain

It is important to review all the data, the routes, and the way of transport, as well as doing continuous improvements of the routes. Recently, air and road freight projects were introduced to improve quality, security and costs, using the minimum number of freight forwarders by managing agreements with global actors.

5.3. Risk Assessment

Airfreights have been proved to be the most challenging process, since the shipment is sent from the site in a truck to a warehouse, where it's packed for airfreight. Then it goes out to the tarmac, and then it goes into the airplane.

When it lands at its destination, we have the same sequence again; standing on the tarmac, into warehouses and then transported to the final destination.

Road transport, however, provides more controlled temperature conditions over the shipment. It is often possible to make door-to-door deliveries.

The risk assessment must consider both quality and security when choosing the most suitable method of shipping the products.

Some freight forwarders have high quality vehicles with GPS tracking systems to locate the position of the trucks along the assigned route, thus increasing the security of the shipment. Also, there are rules for co-loading, multi-drop, and agreed routes to keep the owner of the shipment up-to-date with the progress of his shipment.

Frankfurt, Bangkok, Iceland and Canada, are examples of challenging destinations due to climate and or airport logistics. Especially, the Frankfurt airport could be quite cold during winters and the Bangkok airport is a very hot spot in the summer. Also, both airports have their warehouses quite a distance from the tarmacs, so material is often left in the open places for hours. Considerations on how to improve these aspects are vital. One mitigation could be to book planes that arrive at night in Bangkok to avoid the heat during the day, and the other way around is to ship to Canada and to land during daytime in the winter to avoid the cold nights. Another way to mitigate risks is to protect the shipment by using insulation of different kinds.

6. Shipment Packing Examples

There are several types of packing that can be used to pack the shipment as it is shipped across the globe; for an example, some shipments are put in cardboard boxes and wrapped in polystyrene (see Fig. 21-3). And the other option is to use thermal blankets,which are used as heat insulators to keep the effects of the outside climate to a minimum (see Fig. 21-4).

For products that need to be shipped at a lower temperature, the Envirotainer could be a good choice. It can be transported by air and stays cold with high efficiency using carbon dioxide (see Fig. 21-5).

7. Conclusion

When shipping products on a global basis, it is recommended to be pragmatic and to use risk assessment when setting up a program for temperature monitoring. One should focus on finding the best routes, methods and conditions suiting

Fig. 21-3. Cardboard and polystyrene wrapping

Fig. 21-4. Thermal blanket (1)

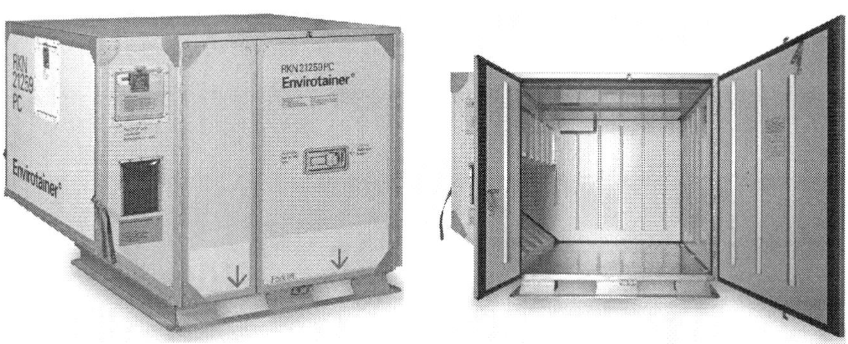

Fig. 21-5. Envirotainer

the product range in question and design the program carefully to manage the complexity of the different markets to meet the overall purpose of securing the quality of the products throughout its lifecycle and avoiding unnecessary product rejections.

It is also important to have easily accessible data so that everybody can review the same data, to have very clear rules and responsibilities on who is going to do what, and to keep good communication between sites. It is necessary to have a system for trending in order to make the necessary adjustments and improvements of the transportation routes and the temperature-monitoring program.

Author Biography

Mr. Conny Axelsson received a B.Sc. in Analytical and Organic Chemistry from Stockholm University. He has worked in the Pharmaceutical Industry for 27 years in the QC and QA area in different positions and leading roles e.g., Director of Quality Assurance and Qualified Person at AstraZeneca Tablet Manufacturing, Gärtuna Sweden. Now in a global role, he is leading and participating in projects and networks with the aim to develop the next generation of Quality Systems within AstraZeneca.

Chapter 22

Introducing a Science-Based Quality by Design Concept to Analytical Methods Development

Jianmei Kochling, Juma Bridgewater, and Redouan Naji

Abstract Although quality by design (QbD) is widely accepted and put into practice for chemical process and formulation development, little attention has been paid to apply it to the analytical methods development and validation. A proposal of applying a science-based QbD approach to a High Performance Liquid Chromatographic (HPLC) analytical method development is presented in this paper. Using the JMP software for DOE design and data analysis, critical method parameters can be identified and thus proper ranges can be set in order to ensure the robustness of the method. This approach allows the robustness to be built into the method during the step of method development.

1. Introduction

A stability-indicating high-pressure liquid chromatographic (HPLC) method, which is designed to monitor the stability of an active pharmaceutical ingredient (API) and its related substances (process impurities and degradants), is usually developed following a systematic approach (Xiao et al. 2007; Li and Rasmussen 2003; Dolan et al. 1998; Jupille et al. 2002). Several software systems, such as Drylab (Rheodyne LLC, CA), ACD (Advanced Chemistry Development, ACD/labs, Toronto, Canada), ChromSword (The ChromSword Group), or automated method development system (AMDS, Waters, MA), are available as tools to assist in method development and optimization. Usually, the focal point is on the method resolution power for each component, and less emphasis has been placed on method robustness during the method development stage.

Concurrent to the quality by design effort on the process development and optimization for chemistry and formulation, the same concept is being implemented for the analytical method development activities. This initiative allows the quality to be built into the analytical methods during the method development stage as opposed to having it tested and improved at the method validation stage.

A quality by design approach for the development of a stability-indicating HPLC method should address issues related to the three essential components:

From: *Pharmaceutical Stability Testing to Support Global Markets*: Pharmasp,
Edited by: K. Huynh-Ba, DOI 10.1007/978-1-4419-0889-6_22,

sensitivity, specificity, and robustness. The method development should thus consist of three steps, (1) establish a method that is capable of separating all peaks of interest with sensitivity by choosing proper wavelength, nominal concentration, columns, mobile phase, etc, following the traditional method development approach; (2) apply a statistical design of experiments approach in conjunction with a confirmative forced-degradation study of the API to assess the robustness and stability indicating power, and to understand which method parameter would potentially cause method failure; and (3) put constraints to the method to ensure the method robustness or further optimize the method.

As an illustration of this concept, this report will provide a case study to describe the process. In order to improve the work efficiency and productivity, statistical design of experiments were employed as a tool for this study.

2. Material

All chemical compounds, API and its related substances, used in this study were made at Vertex. Chemical reagents were purchased from commercial supplier: acetonitrile was from JT Baker; formic acid and trifluoroacetic acid were from Sigma Aldrich. Purified water used was 18.2 MΩ from the Millipore purified water system. A JMP program from SAS was used for the statistical design and the ANOVA program was used for the statistical analysis.

3. Experimental Section

An Agilent 1100 HPLC was used in this study. The HPLC method used a YMC ODS-AQ, 100×4.6 mm, 3.0 μm, 120 Å column (PN# AQ12S031046WT), with mobile phase A composed of 0.025% formic acid and 0.010% trifluoroacetic acid (TFA) in water, and mobile phase B composed of 0.020% formic acid and 0.008% TFA in acetonitrile under gradient elution conditions. The flow rate was 1.5 mL min^{-1} and a column temperature was 35°C. Detection was by ultraviolet (UV) absorbance at 245 nm.

The gradient HPLC method started at 5% B, and ramped up to 90% B in 22 min. The retention time of each related substances is shown in Table 22-1.

Table 22-1. Approximate retention and related retention times of the API and its related substances.

Compound	RT (min)	RRT
Impurity 1	5.9	0.35
Impurity 2	6.5	0.38
Impurity 3	12.4	0.73
Impurity 4	14.7	0.87
API	16.9	1.00
Impurity 5	17.3	1.02
Impurity 6	18.1	1.07
Impurity 7	19.0	1.12

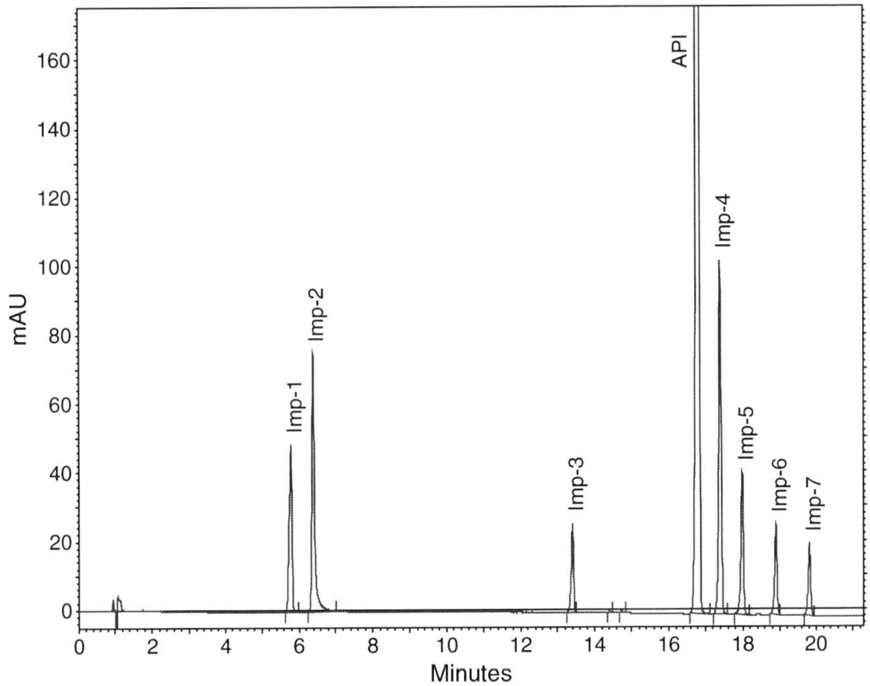

Fig. 22-1. An HPLC chromatogram for the analysis of the API and its related substances spiked at 0.5%, w/w each. The HPLC conditions are listed in the experimental section

4. Results and Discussions

The HPLC method described in the experimental section was developed following column, pH, mobile phase, and wavelength selections. A mixture of API and its related substances (known impurities and degradants) were used for the method development. As shown in the HPLC chromatogram in Fig. 22-1, all peaks were well resolved from one another.

Although the method has shown good separation resolution for each peak, it has not been proven that it would be robust enough for testing from lab to lab, operator to operator. If any of the operation parameters for the method was different from one lab to another, peaks may shift in different directions. Therefore, it would be useful if one can create an experimental design that will include combinational changes of different parameters and perform the experiments to observe the response to these parameter changes.

This work was done using a design of experiments (DOE) approach (JMP software from SAS institute). A set of method parameters were chosen for the DOE experiments, each parameter was set in a continuous mode from low to high values but only tested for the two extremes. As an initial attempt, limited numbers of significant factors (parameters) were chosen, such as organic modifier amount, organic content, flow rate, and temperature, in order to avoid creating a large amount of work and excessive amount of data. A Plackett–Burman design was employed as shown in Table 22-2 (first 5 columns). The Plackett–Burman design is an efficient alternative to a fractional factorial design for screening

Table 22-2. Study results for the 5-factor Plackett–Burman design: relative retention time of each component it the study sample system and assay (potency) for the API.

B (%)	Temp	Flow Rate	FA	TFA	Assay	RRT of Imp 1	RRT of Imp 2	RRT of Imp 3	API	RRT of Imp 4	RRT of Imp 5	RRT of Imp 6	RRT of Imp 7
95	37	1.7	0.01	0	99.76	0.34	0.56	0.88	1.00	1.05	1.07	1.12	1.18
85	33	1.3	0.04	0	99.79	0.35	0.39	0.84	1.00	1.05	1.07	1.13	1.18
85	33	1.7	0.01	0.02	101.03	0.33	0.36	0.70	1.00	1.01	1.07	1.13	1.18
90	35	1.5	0.025	0.01	99.71	0.35	0.38	0.80	1.00	1.04	1.07	1.12	1.18
85	37	1.3	0.01	0.02	99.92	0.35	0.38	0.71	1.00	1.01	1.07	1.12	1.18
85	37	1.3	0.04	0.02	100.42	0.35	0.39	0.70	1.00	1.01	1.07	1.12	1.18
85	37	1.7	0.04	0	100.56	0.32	0.36	0.83	1.00	1.04	1.07	1.12	1.18
95	37	1.7	0.04	0.02	112.26	0.34	0.37	0.70	1.00	1.00	1.07	1.12	1.18
95	33	1.7	0.04	0.02	111.25	0.34	0.38	0.69	1.00	0.99	1.06	1.11	1.17
95	33	1.3	0.04	0	99.99	0.36	0.40	0.84	1.00	1.05	1.07	1.12	1.18
95	37	1.3	0.01	0	99.39	0.36	0.40	0.90	1.00	1.05	1.07	1.12	1.17
85	33	1.7	0.01	0	99.78	0.33	0.37	0.88	1.00	1.05	1.07	1.13	1.18
95	33	1.3	0.01	0.02	100.39	0.36	0.40	0.71	1.00	1.01	1.07	1.12	1.17
Range						0.32–0.36	0.36–0.40	0.71–0.88	1.00	0.99–1.05	1.06–1.07	1.11–1.13	1.17–1.18

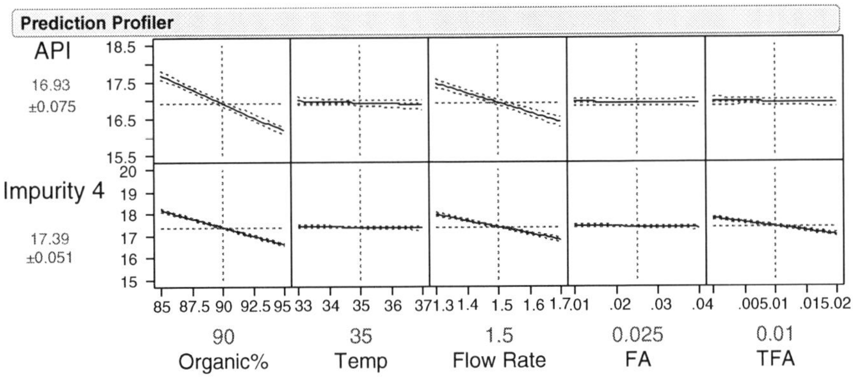

Fig. 22-2. The prediction profiler using the JMP software for the prediction of the retention behaviors of the impurity 4 and the API

the effect of parameters, since the sample size is a multiple of four rather than a power of two, while the sample size for full factorial is 2^n.

Using the design illustrated in Table 22-2, the HPLC analysis results, relative retention time of each related substance (degradant and impurity) and potency of the API, were examined. Ranges of the results of the variation of each parameter were generated. Although the initial parameter variation ranges were very broad, most of the impurity peaks had relatively small relative retention time changes and remained separated. However, a peak labeled impurity 4 merged into the API peak as the flow rate, the organic (B%), and the organic modifier (acids) content were all on the high extremes. The failure may be related to the fact that the initial parameter variation range was too large, for example, the organic% change was 5% as opposed to 2% (a normal range used in method validation).

Taking advantage of the JMP software's predictive profiler feature, which can predict results within the parameter ranges from the existing data when a model was built, the retention behaviors of impurity 4 and API were predicted (shown in Fig. 22-2). The slope representing the retention time change of impurity 4 was slightly lower than that of API while the organic% and the flow rate change. Although it is highly unlikely that all parameters would change simultaneously in one direction, it is nevertheless a hypothetical scenario that can be used to predict when the method failure could happen.

With some knowledge of when the method would fail, a new set of parameters, that would not cause failure, was chosen, and the new relative retention times of impurity 4 and the API were predicted. When setting the parameters to the high extremes for organic (% B) to be at 92% and flow rate to be at 1.7 mL min^{-1}, then varying formic acid% in the range from 0.025 to 0.04% and TFA% from 0.01 to 0.02%, it was found that 30% change in the amount of acid modifiers provided reasonable separation resolution to the API and the impurity 4 (RRT = 1.01–1.02) peaks. Therefore, reasonable parameter ranges that could ensure the robustness of the method are identified, e.g., temperature ±2°C, organic change ±2%, flow rate change ±0.2 mL min^{-1}, and organic acids change ±30%. Given the fact that the probability of having all parameters change at once will be very low in reality, acceptance study results obtained from using this set of newly defined parameters would ensure the method robustness and can be used as normal operation parameter ranges.

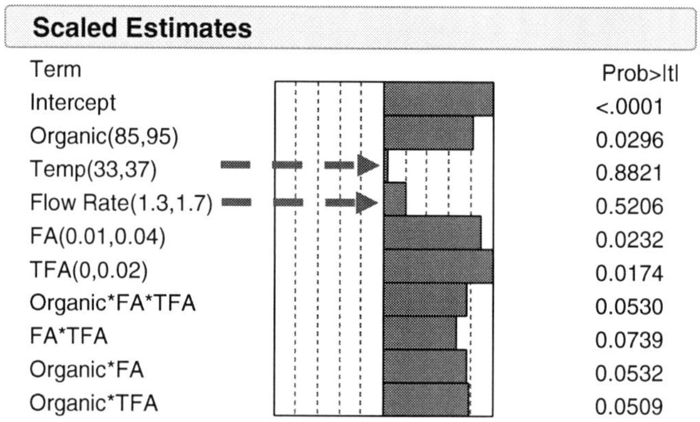

Fig. 22-3. Analysis of the criticality of parameters in the HPLC method using the JMP software. In this case, temperature has no effect and flow rate has minimal effect (*p* \gg 0.05) on the retention time of each component in the sample which consists of the API and its related substances

As for the assay (potency) and retention time of the API, the same strategy was applied. Using the wide parameter ranges mentioned above, when the impurity 4 peak merged into the API peak, the assay was affected and was out of the acceptance criteria range. When the parameter ranges for the organic B% and the acid modifier contents were narrowed, the condition satisfied the separation of impurity 4 and the API. While examining the criticality of the parameters, it was found that the flow rate did not affect the potency and the retention time of the API (Fig. 22-3). Instead, the prediction presented that the other three factors, the content of the organic% from mobile phase B and the two organic acid modifiers were the factors that affected the assay value. However, it may not be true that all three factors were significant, as Plackett–Burman design did allow the recognition of parameter interactions. Thus, a full factorial design was employed for these three factors, organic content, formic acid, and TFA content. Study results showed that only the organic (B%) content affected the assay value significantly, while the two organic acid modifier contents did not influence the API assay value as much. With a good control of the organic (B%) content, e.g., within ±2% change, the potency change was in the range of 98.0–102.0% (w/w), and the change in retention time for API was less than 2%, both were within the acceptance criteria. The experimental data agreed with the DOE software prediction.

A stability-indicating power usually means that this method is capable of separating all impurities and degradants. If the major degradants through early forced-degradation studies have been incorporated into the mixture of related substances, there is no need to rerun the forced degradation samples. However, if standard markers for major degradants are not available, forced-degradation samples should be subjected to the DOE runs along with the API and related substances mixture. This does not imply that all forced-degradation experiments (under all stress conditions) should be repeated. It is important to use the scientific knowledge learned earlier from the characterization of the API, only to repeat those conditions that will cause severe degradations of the compound. In some instances, the API may be extremely and chemically stable, and its

Table 22-3. A 7-factor Plackett–Burman design.

Run	Organic (% B)	TFA (%, w/w)	Formic acid (%, w/w)	Temp (°C)	Flow rate (mL/min)	Column (same brand)	Wavelength (nm)
1	92	0.01	0.0175	33	1.3	b	243
2	88	0.01	0.0175	37	1.7	b	243
3	92	0.005	0.0175	37	1.3	b	247
4	88	0.005	0.0325	33	1.7	b	247
5	92	0.005	0.0325	37	1.7	a	243
6	92	0.01	0.0325	33	1.3	a	247
7	92	0.005	0.0175	33	1.7	a	243
8	90	0.0075	0.025	35	1.5	a	245
9	92	0.01	0.0325	37	1.7	b	247
10	88	0.005	0.0325	33	1.3	b	243
11	88	0.01	0.0325	37	1.3	a	243
12	90	0.0075	0.025	35	1.5	b	245
13	88	0.01	0.0175	33	1.7	a	247
14	88	0.005	0.0175	37	1.3	a	247

stability is not influenced by pH. In other instances, the compound is extremely unstable, in which case, the major degradants should be identified and marker standards should be available to be mixed for the initial DOE studies. In this case study, only oxidation was repeated, since the compound is a very stable compound. Using the newly identified robust operational parameter ranges, Organic B% = 90 ± 2%, Flow Rate = 1.5 ± 0.2 mL min^{-1}, TFA% = 0.007 ± 30%, and FA% = 0.025 ± 30% along with two more method parameters, wavelength and column were used for another Plackett–Burman DOE study (Table 22-3). A mixture that consisted of degradants from the oxidative forced degradation, API, and process impurities was used as the investigational sample system. One thing worth pointing out was that the initial DOE results suggested that too much of TFA would cause the impurity 4 to move to shorter retention time and closer to the API peak (Fig. 22-2). Thus, the amount of TFA was adjusted to be 30% of the original, and this also helped the baseline to be flatter and thus easier for the impurity peak integration. The results (Tables 22-4 and 22-5) for the relative retention times and USP resolution between two adjacent peaks showed that even when there was an overlap, (for Deg-1/Imp-1 and Imp-1/Imp-2) it resulted from the summary of all runs, in each run, the resolutions between two adjacent peaks were all >3. Thus, the method is robust to separate all peaks within the parameter variation range and it is a stability-indicating as the chromatogram shown in Fig. 22-4.

In a hypotentical case, if the forced degradation peaks moved too close to the known related substances, one should consider to modify the existing HPLC method since the method is still being developed. The prediction profiler from JMP can be used as an experimental guide for the selection of the starting point of the experimental conditions.

The study case mentioned above introduced a concept which can be depicted by Fig. 22-5. At the bottom of the pyramid where the parameter variation range is the largest, the responses (potency or retention time values) have

Table 22-4. Results obtained for the 7-factor Plackett–Burman design using a sample mixture which consists of the oxidative stressed mixture, the API, and the process impurities.

Run	Deg-1	Imp 1	Imp 2	Imp 3	Deg-2	Imp 4	Imp 5	Imp 6	Deg-3	Imp 7	Imp 8
1	0.28	0.33	0.37	0.74	0.83	0.86	1.03	1.07	1.10	1.13	1.18
2	0.24	0.29	0.33	0.73	0.82	0.85	1.03	1.07	1.10	1.13	1.19
3	0.24	0.29	0.33	0.72	0.82	0.86	1.02	1.07	1.10	1.13	1.19
4	0.27	0.32	0.36	0.79	0.83	0.86	1.04	1.07	1.10	1.13	1.18
5	0.29	0.34	0.38	0.78	0.84	0.87	1.03	1.07	1.10	1.12	1.18
6	0.31	0.35	0.39	0.80	0.84	0.86	1.04	1.07	1.10	1.12	1.17
7	0.24	0.29	0.33	0.76	0.82	0.86	1.04	1.07	1.10	1.13	1.19
8	0.24	0.29	0.33	0.77	0.82	0.85	1.03	1.07	1.10	1.13	1.19
9	0.27	0.32	0.36	0.77	0.83	0.86	1.04	1.07	1.10	1.13	1.18
10	0.28	0.33	0.37	0.73	0.83	0.86	1.03	1.07	1.11	1.13	1.18
11	0.24	0.29	0.33	0.72	0.82	0.85	1.02	1.07	1.10	1.13	1.19
12	0.27	0.32	0.36	0.73	0.83	0.86	1.03	1.07	1.10	1.13	1.18
13	0.26	0.31	0.35	0.75	0.83	0.86	1.03	1.07	1.10	1.13	1.19
14	0.26	0.31	0.35	0.75	0.83	0.86	1.03	1.07	1.10	1.13	1.19
Range	0.24–0.31	0.29–0.35	0.33–0.39	0.72–0.80	0.82–0.84	0.85–0.87	1.02–1.04	1.07	1.10	1.12–1.13	1.17–1.19

Table 22-5. USP resolutions of two adjacent peaks in the chromatograms obtained for the 7-factor Plackett–Burman design using a sample mixture which consists of the oxidative stressed mixture, the API, and the process impurities.

Run	Deg-1/ Imp 1	Imp 1/ Imp 2	Imp 2/ Imp 3	Deg-2/ Imp 3	Imp 4/ Deg 2	API/ Imp 4	Imp 5/ API	Imp 6/ Imp 5	Deg-3/ Imp 6	Imp 7/ Deg-3	Imp 7/ Imp 8
1	5.6	3.6	47.2	15.0	5.0	21.0	4.2	5.7	4.1	3.6	7.5
2	6.5	3.6	51.6	14.8	4.9	22.5	3.6	6.7	3.8	3.9	8.1
3	6.6	3.8	46.4	16.6	5.2	22.5	3.6	6.9	4.1	4.1	8.2
4	5.8	3.3	51.2	5.8	4.6	21.7	5.6	4.4	4.0	3.4	7.7
5	6.2	3.8	52.3	8.3	5.1	22.1	5.3	5.1	4.1	4.0	8.2
6	6.3	3.5	53.0	6.1	4.8	22.5	5.7	4.7	4.1	3.6	8.0
7	6.0	3.9	53.6	10.3	5.1	22.0	4.9	5.3	3.9	4.0	8.1
8	6.3	3.3	49.4	8.5	4.8	22.3	4.9	5.3	3.7	3.7	8.1
9	6.4	4.2	52.3	9.3	5.0	21.6	5.6	4.6	4.2	3.8	7.7
10	5.8	3.9	46.3	16.5	4.9	20.9	4.0	6.0	4.2	3.6	7.5
11	6.1	3.4	49.9	16.2	4.9	22.6	3.3	6.9	3.8	3.7	8.2
12	5.6	3.2	46.5	15.2	4.6	21.9	3.9	6.1	4.0	3.5	7.8
13	6.3	3.6	47.9	12.8	5.0	22.2	4.7	5.8	4.1	3.9	8.1
14	6.0	3.6	50.7	12.7	5.0	22.1	4.5	5.7	4.0	3.8	7.9

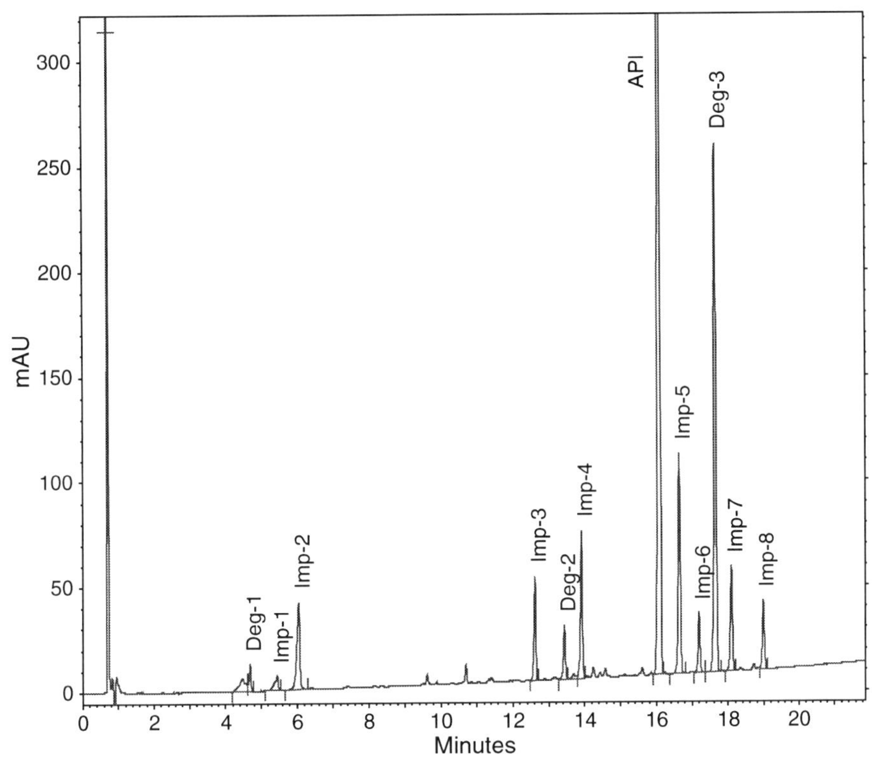

Fig. 22-4. An HPLC chromatogram obtained using a mixture sample that consists of the oxidative forced-degradation mixture, the API, and the process impurities

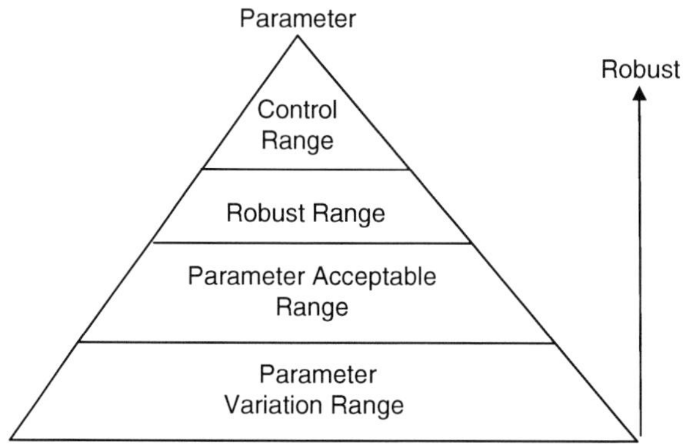

Fig. 22-5. A diagram that depicts the relationship between the parameter variation ranges and the method robustness

the largest variation and the method is the least robust. As the parameter variation range is getting smaller, eventually to the robust range and then to the controlled operation range, the method should be robust under the set of operation parameters. This method development strategy not only allows quality to be incorporated into the method, it also allows the scientists to have a thorough understanding of the method variability in relationship to the method parameters.

5. Conclusions

Unlike the conventional method development approach, a new concept using statistical design of experiments has been introduced to the development of a stability-indicating HPLC method. This approach allows the critical parameters for the HPLC method to be assessed in a systematic way, resulting in a thorough understanding of the variability of the method in association with parameter changes. Additional benefit obtained from this approach is that a large amount of scientific information is generated with a relatively small amount of laboratory effort, which directs scientists to see if the method needs to be further optimized. In order to accomplish the above mentioned tasks, selection of a proper surrogate sample system that consists of impurities and potential degradants is essential. Using the scientific quality-by-design method development strategy, the method robustness is built into the method during the method development step and this ensures that the method can be used for long term.

References

Dolan JW, Snyder LR, Djordjevic NM, Hill DW, Saunders DL, Van Heukelem L, Waeghe TJ (1998) J Chromatogr A 803:1–31

Jupille TH, Dolan JW, Snyder LR, Molnar I (2002) J Chromatogr A 948:35–41

Li W, Rasmussen HT (2003) J Chromatogr 1016:165–180

Xiao KP, Xiong Y, Liu FZ, Rustum AM (2007) J Chromatogr A 1163:145–156

Author Biographies

Dr. Jianmei Kochling received her Ph.D. degree from Northeastern University in analytical chemistry. Dr. Kochling is currently an Associate Director of analytical R&D at Genzyme Corporation. She leads the analytical teams in support of CMC activities, chemical process, and formulation development for small molecule development programs.

Dr. Juma Bridgewater received his Ph.D. degree from University of Massachusetts, Ammherst in analytical chemistry. Dr. Bridgewater is currently a Scientist at Abbott Bioresearch Center in Worcester, MA. He performs preformulation studies in support of small molecule discovery programs and antibody development programs.

Mr. Redouan Naji received his B.S. degree from Institute Notre Dame, Belgium in Biochemistry. Redouan is currently a Senior Scientific Associate at Vertex Pharmaceuticals in Cambridge, MA. He performs Analytical method developments in support of small molecule development programs.

Section IV

Stability Data and Operational Practices

Chapter 23

Optimizing Stability Data Package to Facilitate NDA/MAA Approval

Frank Diana

Abstract The stability data package is often the last piece completed prior to finalization of the Chemistry, Manufacturing and Controls (CMC) section of an NDA/MAA regulatory submission. The data contained in the stability section is used to determine the proposed drug product expiration date and API retest date, to establish any special requirements for packaging, to facilitate justification of the API and product specifications, as well as to provide commitments for the commercial stability program. Well planned and executed stability studies, as well as an accurate evaluation, and well thought out presentation of the data will aid regulatory review. This paper focuses on the stability studies required to meet the current regulatory guidelines with an emphasis on the design of the studies and the evaluation, interpretation and presentation of the data.

1. Introduction

In today's environment, the ICH guidelines are typically used to structure the stability program for an upcoming NDA/MAA submission. Approximately 18–24 months before the scheduled filing date, plans should be underway for the primary stability batches, e.g., the final formulation and packaging, number of strengths, manufacturing site, filing strategy (US only, Global, US/Europe, etc). The ICH parent stability guideline, Q1A, provides the requirements for a submission in climatic zones 1 and 2, which covers US, Europe and Japan and many other countries. If a submission is to be filed in alternate regions, additional guidelines should be consulted and requirements included in the stability program as necessary, e.g., additional storage conditions, more protective packaging and testing requirements. In addition to ICH Q1A, other relevant stability guidelines should be reviewed and appropriate requirements should be included in the stability program, i.e., photostability (Q1B), bracketing and matrixing (Q1D), stability data evaluation (Q1E). Additional studies will be

From: *Pharmaceutical Stability Testing to Support Global Markets*: Pharmasp,
Edited by: K. Huynh-Ba, DOI 10.1007/978-1-4419-0889-6_23,

included in the stability program such as bulk holding, shipping, thermal and In-Use studies as needed, consistent with the type of dosage form and submission/ commercialization strategy. Based on an evaluation of the data generated for the primary stability batches, an expiration date for the drug product and a retest date for the API will be proposed. Drug Product and API specifications will be finalized and justified for submission utilizing the data from primary stability studies, as well as developmental studies and other information including regional requirements. Depending on the scale of the primary stability batches, a stability commitment and post-approval stability protocol will be written and included in the submission. The post-approval stability protocols will include final testing requirements and packages based on the results from the primary stability studies and may include bracketing/matrixing designs for the commitment batches, if justified. The justification may include deletion of non-value added tests for the post-approval stability protocols based on the data generated. And to address future potential post-approval product changes, a comparability protocol may be included in the submission. The submission requirements for these protocols are also presented.

2. ICH Q1A Requirements

The ICH Q1A guidance document describes the stability requirements for the submission packages of both new drug substances and new drug products. The document provides direction on the number of batches to be manufactured following the same process to be used for commercial production, as well as the scale of the batches required. The batches placed on stability should be tested for those attributes that are susceptible to change over time, and include physical, chemical, biological and microbiological attributes, preservative content and functionality tests as appropriate. The storage conditions (for API and drug product intended to be stored at room temperature), time points and expected data to be included in the filing are shown in Table 23-1.

Samples are typically stored at Intermediate conditions (30°C/65% RH) but not tested unless there is a significant change at the accelerated condition (40°C/75% RH). A significant change for an API on stability is defined as a failure to meet any of the API specifications. For a drug product on stability, a significant change is defined as:

- 5% change from initial assay value.

Table 23-1. ICH requirements for a new drug product or API.

Storage conditions[a]	Time points	Available data for submission
25°C/60% RH (controlled room temperature)	0, 3, 6, 9, 12, 18, 24 months and yearly thereafter to desired expiration or retest period	12 months
30°C/65% RH (intermediate)	0, 6, 9, 12 months	6 months from an ongoing 12 month study
40°C/75% RH (accelerated)	3 time points, e.g., 0, 3, 6 months	6 months

[a] Stability chambers controlled at ±2°C and ±5% RH

- Any degradation product exceeding its specification limit.
- Product exceeding pH limits.
- Dissolution exceeding the acceptance criteria for 12 units.
- Failure to meet specifications for appearance and physical properties.

If a significant change is observed, the intermediate condition study is typically picked up at this point with a minimum requirement of 4 time points as shown in Table 23-1. For the drug product stored at accelerated conditions, a 4th time point should be added if significant change is approached or additional samples can be tested at 6 months to assure appropriate assessment as to whether significant change criteria are met. Based on the evaluation of the stability data generated at the different storage conditions, the API retest period or the drug product's expiration date will be proposed as described below.

3. ICH Requirements: Alternate Stability Storage Conditions

3.1. Semi-Permeable Containers

For aqueous based products, such as oral solutions, opthalmics, nasal sprays, large and small volume parenterals, packaged in semi-permeable containers, such as plastic bottles, bags and vials, the relative humidity settings of the stability storage conditions are different in order to challenge the containers with regard to water loss. Since semi-permeable containers have the capability to transmit water vapor through the walls of the container, the relative humidity requirements are lower for these types of containers in order to present a worst-case type condition. The storage conditions for these products are long term: 25°C/40% RH, accelerated: 40°C/NMT 25% RH, and the intermediate condition remains 30°C/65% RH. For a global product, the long term condition may be 30°C/35% RH, in which case there is no intermediate storage condition. In addition to the definition provided above for drug products, a significant change in water loss is defined as a 5% loss after 3 months at 40°C/NMT 25% RH, although a significant change in water loss alone does not necessitate the testing of samples storage at intermediate conditions. If stability chambers are not available at these alternate relative humidity conditions, samples can be stored in the typical long term and accelerated conditions described in Table 23-1, and the water loss can be derived at the reduced relative humidity by calculation. For example, take the RH at normal accelerated conditions, 75%, and divide by the reduced RH of 25% to obtain a factor of 3; multiply the water loss determined at 40°C/75% RH by 3 to estimate the water loss at 40°C/25% RH.

3.2. Refrigerated Conditions

For drug products labeled for refrigerated storage, the long term storage condition is 5°C±3°C with the accelerated conditions being 25°C/60% RH and no intermediate condition. The accelerated condition is a compromise so as not to require another stability chamber (e.g., 20°C). Drug products are typically labeled for refrigerated conditions due to instability at controlled room temperature; therefore, it is not unusual to find a significant change at

25°C/60% RH. If this change occurs between 3 and 6 months, then the proposed shelf life will be based on the available real time data at 2–8°C with no extrapolation. If the significant change occurs in the first 3 months, then data to cover use (e.g., supply chain handling, shipping, emergency vehicles) of product outside of label storage condition must be supplied. It is not necessary to test at 6 months accelerated storage when an obvious significant change occurs in 3 months.

4. Reduced Stability Testing Design

The ICH Q1A document indicates that batches for each proposed strength and container size of the drug product should be placed on stability and tested, unless a matrixing and bracketing design is used to reduce the testing frequency, and justification for the design is provided. To provide additional guidance on these types of study designs, ICH Q1D *Bracketing and Matrixing Designs for New Drug Substances and Products*, was issued in 2003. Bracketing is the design of a stability schedule so that only the samples on the extremes, for example of container size and/or dosage strengths, are tested. Bracketing is applicable to most types of products and can be justified for formulations that are very closely related, such as dosage strengths containing different colorants or flavors. This type of study design is particularly advantageous for tablet products, which are manufactured using a common granulation across several strengths. To illustrate this, consider an immediate release product available in 25, 50, 100 and 200 mg strengths prepared from the same blend (tablet weight is proportionately increased with each strength). Using bracketing concepts, only the extremes would be placed on stability, in this case it would be of the 25 and 200 mg strengths. The intermediate strengths would be covered by the stability data generated for the 25 and 200 mg dosage forms. A similar approach can be used for container sizes, e.g., 40 cc, 75 cc, and 325 cc HDPE bottles, in which the components are the same. In this case, the tablet to volume ratio would need to be taken into account to determine the extremes. For this example, the 40 cc and 325 cc bottles will represent the extremes. Taking the strength and container size factors together, 12 product combinations (4 strengths × 3 package sizes) would be supported by the 4 product configurations representing the extremes; 25 mg/40 cc, 25 mg/325 cc, 200 mg/40 cc and 200 mg/325 cc (see Table 23-2). For the primary stability studies, which require 3 batches per configuration, a total of 12 batches would be placed on stability instead of 36. This example illustrates that a good bracketing design can save a substantial amount of stability resources and assures that only value-added testing is performed.

Table 23-2. Bracketing example (X = sample configuration stability).

Package/strength	25 (mg)	50 (mg)	100 (mg)	200 (mg)
40 cc HDPE bottle	X	–	–	X
75 cc HDPE bottle	–	–	–	–
325 cc HDPE bottle	X	–	–	X

Matrixing is defined as the statistical design of a stability schedule so that only a fraction of the total number of samples is tested at any specific sampling point. Factors that can be matrixed include batches, strengths, container/fill sizes, and intermediate time points. The initial and final time points for each sample need to be tested; for primary stability batches, this translates to testing at time zero and 12 months for the 25°C/60% samples assuming the 12 month data are the latest data that will be available at the time of filing. Matrixing designs should typically be reviewed with regulatory authorities (e.g., end of phase II meeting) and need to be supported by available data. The less variable the available stability data and the more stable the product, the more likely the matrixing design will be justified. An example of a matrixing design is provided for an extended release product developed in two strengths and packaged in two HDPE bottle sizes (e.g. 4 combinations × 3 batches of each = 12 primary stability studies). In this case, a bracketing approach would not be appropriate since there are not enough combinations to establish extreme/intermediate samples. However with the appropriate data as described above, a matrixing design can be justified to test only a fraction of the samples that would be tested in a full stability protocol design. For example, applying matrixing (testing ½ the samples) to the primary stability samples, the design would be as follows:

Full protocol – 0, 3, 6, 9, 12, 18, 24, 30 and 36 months

Matrixing – all batches 0, 12, and 36 months and then ½ of the intermediate time points tested = six samples per batch

Using this approach, the testing has been reduced from 108 samples (12 batches × 9 time points) to 72 samples.

Bracketing and matrixing can also be combined where appropriate, e.g., same case as above, but add a third similar package which meets bracketing criteria. As illustrated by these examples, matrixing and bracketing alone or combined can save a significant amount of stability testing resources and substantial time in preparing to file a NDA/MAA. The post-approval stability protocols for commitment batches (first three production batches) and annual batches can also utilize these same concepts, thus multiplying the resource savings over the product lifetime.

5. Testing Considerations

When planning and then executing stability studies, numerous testing factors need to be taken into account over and above the analytical method used, the time points tested and the storage conditions. For liquid products, container orientation must be included to determine whether there is any difference between upright and inverted storage. Container orientation evaluation is typically included for the primary stability batches and if no difference is observed, typically the worst-case orientation is used going forward, e.g., inverted. Container extractables including label components (adhesive, ink) are important tests performed as part of packaging development to assure the packaging is not contributing unacceptable levels to the product. Container/closure systems can also impact product stability through absorption of the drug substance or preservatives, and thus, assay of these ingredients during primary stability is critical for the appropriate dosage forms. Products that contain preservatives need to meet antimicrobial preservative effectiveness test (AET) requirements (e.g., USP<51>). During development, the effectiveness of the preservative system can be challenged to show that at lower preservative levels (i.e., with samples formulated with preservative at 50–80%) the system is still effective. If the preservatives are shown to be effective at these lower levels, justification for a lower specification limit will be supported as well as replacement of this time consuming test with assay of the preservatives during stability testing.

Semi-solid dosage forms packaged in multiple use containers should be tested for homogeneity (top, middle, bottom of container) throughout the primary stability studies to assure that there is no migration or separation of the components or inconsistent degradation depending on the location in the container.

For all types of dosage forms, emphasis should be placed on reporting of degradation products during primary stability studies. Identification and/or qualification of degradation products should also be undertaken based on the criteria presented in ICH Q3A and Q3B (Impurities in New Drug Substances and New Drug Products, respectively) prior to the submission of the NDA/MAA. As part of stability testing, emphasis should be placed on the determination of mass balance throughout the study. ICH Q1A defines mass balance as "*The process of adding together the assay value and levels of degradation products to see how closely these add up to 100% of the initial value, with due consideration of the margin of analytical error.*" Consistently poor mass balance or a decreasing trend may be indicative of a method or product problem or both, and needs to be investigated and resolved. If this issue is left unresolved, it could indicate that there are unknown degradation products being formed, which are not accounted for, and ultimately this could impact product approval.

6. ICH Requirements: Photostability

ICH Q1B Guidance Document describes the process for determining if a new drug substance or product is sensitive to light so that the proper precautions such as protection during production or protective packaging can be taken to promote stability. During photostability studies, samples are exposed to light providing an overall illumination of not less than 1.2 million lux hours and an integrated near UV light energy of 200 watt hours/square meter or more. The ICH guideline includes a flow diagram to guide the steps to be taken to complete this study. Initially, API is exposed directly or in solution as part of the forced degradation studies necessary for analytical method development and validation. Subsequently, the API (typically, one batch) is subject to confirmatory studies to determine if any special handling, packaging or labeling is necessary to ensure the stability of the material. Samples are typically stored in glass dishes and protected with a transparent cover, if necessary. Analysis of the samples is performed to determine if any physical or chemical changes have occurred indicating that special precautions are necessary with regard to light exposure.

For drug product, confirmatory testing is carried out sequentially starting with directly exposed product, continuing with the product in its immediate package, and finally with product in the marketing package if necessary for protection from light. Typically one batch of each formulation is studied and analyzed, and if after direct exposure acceptable change is noted physically or chemically, the study is ended. On the other hand, if there is an unacceptable change such as a significant decrease in assay or increase in degradation products, then testing will continue with the immediate container/closure system to determine if it provides adequate protection. The same evaluation criteria are applied, and if needed, the final market package (secondary packaging) is tested. After the completion of all steps, if the product still has an unacceptable change, redesigning of the package or reformulation may be necessary. Depending on the outcome of the studies, labeling statements such as "Protect from Light" may be necessary, and in those cases where secondary packaging is necessary, an additional statement such as "Retain in carton until contents are used" may be warranted.

7. Additional Studies

Depending on the nature of the product and its manufacturing process, additional studies should be considered as appropriate to ensure the stability profile of the product is fully characterized. While not necessarily required for submission, these studies should be performed to provide the manufacturer the information needed to ensure the stability of the product.

Bulk Product Studies: If a product is to be stored in the manufacturer's warehouse for a period of time between production and packaging, a bulk stability study should be conducted to support that storage period. The study should be carried out under controlled conditions in a stability chamber using a simulated package such as a small fiber drum (with plastic bags) or a plastic container that represents how the bulk product will be stored. If this is not possible, a study in the actual bulk container in the warehouse may be acceptable with monitoring of the storage conditions throughout the study and subsequent calculation of the mean kinetic temperature. This study can be done on a non-commercial batch as long as it accurately represents commercial manufacturing and storage. Stability indicating methods must be used to test the product and appropriate acceptance criteria must be used to assure that the product packaged after bulk storage would meet specifications through its shelf life.

In-Process/Intermediate Materials: For materials known to be stable, a holding period of 30 days under appropriate storage conditions is generally acceptable without conducting stability studies to verify the holding periods. However, for unstable materials or for materials held longer than 30 days before use in manufacturing a finished dosage form, stability studies are necessary to verify the holding periods for the in-process / intermediate material. These studies are typically combined into a finished product stability study so that first the in-process stability is established, and then the product manufactured from the held material is placed on stability. In addition, bulk product holding time can be added to this study to generate stability data for the entire manufacturing and packaging cycle; e.g., in-process material held and tested, bulk product manufactured, held and tested, product packaged and placed on stability through its shelf life.

Shipping/Thermal Studies: If the product is to be manufactured at one site and then shipped to another site for packaging, stability studies need to be performed to support storage and shipment of the bulk container, as well as placing the final packaged product on stability as representative of the supply chain. These stability studies would take into consideration the typical temperature patterns experienced at different times of the year during shipping and would confirm that the bulk product is not affected physically or chemically.

Thermal studies would typically be performed on drug products susceptible to phase separation, aggregation, precipitation and/or viscosity changes. These studies would support shipping and handling throughout the supply chain. Such studies involve cycling over a range of temperatures that the product may be expected to encounter during distribution, for example, three cycles of 2 days at −20°C and 2 days at 40°C. The actual temperatures used for the cycling should take into consideration any labeling statement. For example, for products labeled with a "Do Not Freeze" statement, the temperature cycling may be conducted between 2–8°C and 40°C. During these studies, it is important to evaluate any changes that occur in order to determine if the product is impacted. For example, an ingredient such as a preservative may precipitate out of solution at cold temperatures, however on warming it may re-dissolve. It is important to understand the potential outcomes and establish appropriate shipping and storage requirements.

In-Use Stability: The expectations for this type of study are documented in European Guidance CPMP/QWP/2934/99. The objective is to evaluate the time a multidose product can be used once the container is opened. The study should simulate the use of the product in practice, e.g., dilution/reconstitution time, sampling under normal environmental conditions, storage as specified on the label after dilution or reconstitution. The test should be performed to cover the shelf life of the product, which may be difficult for a new product before filing; however the test can be performed at 12 months to determine if any changes have occurred on storage. The appropriate physical, chemical and microbial testing should be performed based on the dosage form and focusing on those parameters that can change after a multidose container is opened.

8. Specifications

Data from the primary stability batches are considered in the establishment and justification of specifications for the new API or new drug product, along with the data from development batches, batches used in clinical and toxicology studies (impurity/degradation product qualification), as well as reasonable analytical and manufacturing variability. Trending of stability data during development facilitates understanding of the critical issues and establishment of appropriate alert and in-house limits for marketed batches to assure that the released product will meet specifications throughout shelf-life and that out of profile results on stability will be investigated. Appropriately established release requirements can help to avoid typical issues that arise on stability. Some examples of release testing issues that can potentially show up on stability include:

- High initial degradation product (impurity) level in API.
- Atypical variability in tablet assays (poor content uniformity, segregation of active ingredient).
- pH near limits at release (analytical variability).
- Stage 2 dissolution, or atypically low average dissolution result.

In each example, specifications are met but the risk of problems on stability are not considered or not deemed critical. These issues could be better understood upfront and evaluated proactively with the appropriate release requirements. Guidance on establishing specifications is provided in ICH Guidance Document Q6A *Specifications: Test Procedures and Acceptance Criteria for New Drug Substances and New Drug Products: Chemical Substances.*

9. Stability Data Evaluation and Expiration Dating

ICH Guidance Document Q1E describes how to determine a proposed expiration date or retest period and includes direction on when limited extrapolation of the data may be acceptable. A minimum of three batches should be evaluated and the shelf life estimated based on where the 95% one-sided confidence limit intersects the specification limit. The guideline also discusses the subject of poolability with the most important outcome being that if samples cannot be statistically pooled, then the overall shelf life will be set based on the shortest shelf life estimated (worst case). Appendix A of the guidance provides a very useful decision tree for data evaluation and limited extrapolation (3–12 months) of retest or expiration dating periods for drug substance and product. Factors that may impact extrapolation include whether significant change occurs at accelerated or intermediate conditions, what type of change and variability is observed for the long term data, whether or not the data are amenable to a statistical analysis and whether one has been performed. In the best case, extrapolation of twice the available long term data (not exceeding 12 months) may be appropriate. For example, extrapolation of 12 months long term data to 24 months retest or shelf life may be appropriate when no significant change has occurred through 6 months at accelerated conditions, the long term data indicates a stable product with little data variability and the intended storage condition is controlled room temperature. Extending this example, if 18 months long term data are filed during the review period and the data remains acceptable, a proposed shelf life of 30 months would be appropriate. If a product is to be stored at refrigerated conditions, an extrapolation of only 1.5× is allowed; therefore, 12 months long term data would support a proposed shelf life of 18 months. Where significant change is observed within 6 months at the accelerated conditions and at intermediate conditions, extrapolation is not permitted. Similarly, for refrigerated products, if significant change is observed within 3 months at accelerated conditions (25°C/60% RH), no extrapolation is warranted.

The proposed period of expiration date at the intended storage conditions will be included in the submission based on the evaluation of the available stability data. With the approval of the NDA, the expiration date is considered tentative until confirmed by sufficient long term stability data. The post-approval and/or primary batch stability protocol included in the NDA should include time points past the proposed expiration date if extension is desired in the future. Extension of the expiration date can be done via two regulatory reporting options: an annual report update or a prior approval supplement. The mode of choosing the usage option depends on the scale of the batches used to support the request. When the expiration dating extension is based on three production scale batches completing stability through the desired expiration date, the annual report may be used. These batches can be the original primary stability batches if made at production scale or the post-approval commitment production scale batches. For example, if the original expiration date granted at the time of NDA approval is 18 months and the approved protocol filed in the NDA included testing through 36 months, once the three production scale batches reach 24 months with acceptable results, the product expiration date can be updated to 24 months. Similarly, once these batches reach 36 months with acceptable results, the expiration date can be extended to 36 months in the annual report. The second approach is to use the original registration batches, which are often manufactured at pilot scale. In this case, once the pilot batches are tested at the desired expiration date under an approved protocol with acceptable results, a prior approval supplement can be submitted proposing extension of the expiration date. The applicant would then wait for FDA approval of the shelf life extension before changing the expiration date for subsequent manufactured/packaged product.

10. ICH Requirements: Stability Commitment

The stability commitment and post-approval stability protocols have been noted in previous sections. A stability commitment needs to be included in the NDA/MAA and ICH Q1A clarifies the production scale batch stability requirements as follows:

- If the submission includes long-term storage data from:
- Three production batches, commitment to continue studies through proposed shelf life
- Fewer than three production batches, commitment to add production batches to at least three on long-term and accelerated stability through proposed shelf life
- No production batches, commitment to place first three production batches on long-term and accelerated stability through proposed shelf life

ICH Q1A also indicates that the stability protocol for commitment batches must be the same as for the primary submission batches, unless justified otherwise. It goes on to clarify that;

- If significant change occurs on primary batches at accelerated condition:
- Commitment batches should be tested at intermediate condition instead of accelerated condition.
- Alternatively, testing can be done for samples stored at accelerated condition for commitment batches, but if there is any significant change, test samples stored at intermediate condition.

11. Comparability Protocols

Comparability protocols are described in a draft FDA guidance. The inclusion of a comparability protocol in the original NDA can be very useful for certain post-approval changes. The protocol prospectively specifies tests and studies including stability testing to be performed based on the type of change to be made. The corresponding acceptance criteria are also included in the protocol. In the NDA, a sponsor may include a comparability protocol for an anticipated change, such as a package change or a manufacturing site change or an API synthesis change, and describe the testing/studies that will be completed to qualify the change prior to implementation. Typically, product and API specifications would remain unchanged in this type of protocol. The advantage to the applicant in filing a comparability protocol is to request that the FDA allows the specified change to be reported at one category lower than the normal. If the FDA agrees then, for example, a change that is typically filed as a CBE supplement may instead be filed in the annual report. Effort put into designing quality into the product through quality by design principles and utilizing process analytical technology tools during pharmaceutical development will facilitate this approach.

12. Stability Data Package

Taking into account all the studies recommended above, a substantial amount of information should now be available to include in the NDA/MAA stability section, as well as facilitate preparation for regulatory agency inspections and validation/launch of commercial product. Summarizing the information in a thorough and well thought out manner, presenting the data in an easy to follow format, and focusing attention on the critical points related to stability is the key to facilitating review of this section by the regulatory authorities. The package should include:

- Stability protocols for the primary and supporting batches.
- Significant deviations/investigations.
- Statistical Analysis, if appropriate.
- Proposed expiration & retest dates.
- Label storage conditions.
- Post-approval commitments & protocols.
- Data tables including tests and limits, relevant batch information (API, manufacturing date & site, packaging, etc.).
- Stability (SAS) transport files.

Since the stability package is often the last section to be included in the CMC portion of the NDA/MAA, effective planning is needed to complete all studies as needed and evaluate the data in a timely manner. Often, this package is assembled and reviewed before the final time points are pulled and tested. Then once the final data are available, the necessary information is updated (data tables, statistical analysis, summary, proposed expiration date) and sent for a final review and approval before incorporation into the final CMC section. An optimized stability package with well designed studies and thorough analysis and presentation of the information will help to ensure the stability of the product and approval of the submission.

References

ICH Q1A(RA2) – Stability testing of new drug substances and products
ICHQ1B – Photostability testing of new drug substances and products
ICH Q1D – Bracketing and matrixing designs for stability testing of new drug substances and products
ICH Q1E – Evaluation of stability data
ICH Q3A(R) – Impurities in new drug substances
ICH Q3B(R) – Impurities in new drug products
ICH Q6A – Specifications: test procedures and acceptance criteria for new drug substances and new drug products: chemical substances
CPMP/QWP/2934/99 – In-use stability testing
Guidance for Industry: Comparability Protocols – chemistry, manufacturing and controls information, Draft Guidance 02/2003

Author Biography

Dr. Frank J. Diana is a Vice President of Pharmaceutical Development for Endo Pharmaceuticals, Chadds Ford, PA. Previously, he was senior director of Technical Operations at Endo responsible for analytical support for marketed products as well as technology transfer, scale-up, and validation of new products; and director of QA New Products at J&J, Titusville, NJ responsible for QA activities related to the introduction and transfer of new products into manufacturing. Currently he is responsible for formulation, analytical and clinical supply activities for the Endo development pipeline as well as providing support to business development and life cycle management projects. He previously worked for 25 years at DuPont Pharmaceuticals, where his last position was director of analytical technology, which included chairing the Pharmaceuticals Stability Review Board and Validation Review Board. A member of AAPS, ACS, ISPE, PSDG and the PhRMA Stability Expert Team, he coauthored the ISPE Guideline for Technology Transfer of Analytical Methods. He is an adjunct professor in the QA/RA graduate program at Temple University's School of Pharmacy.

Chapter 24

Maximize Data for Post Approval Changes

Paula J. Youngberg Webb

Abstract During the life of a drug product, it is almost certain that at some point, changes will be made, for example to the manufacturing process, the API material, or the formulation. For each change, it is necessary to demonstrate the acceptability of the proposed change, in order to prove that the specified change does not have an adverse effect on the product. The type of change will determine what type of stability data will be required to support it and to understand the impact on the stability profile of the drug product. An appropriate stability design must be developed to understand the impact of the change on the stability profile of the drug product and to demonstrate the acceptability of the proposed change. The type of change will dictate the amount of data needed and the type of regulatory reporting required. Guidance documents provide direction on the reporting category for different types of changes. For a proposed study design or data package that is different from the guidance documents, it is best to discuss the proposal with the regulatory agency first to obtain their opinion. This presentation will review the various reporting categories and provide stability strategies to support some specific types of changes, including some case examples of stability designs that have been conducted.

1. Introduction

Once a product is commercialized, at some point in its life cycle, something will change. Products rarely stay the same. It may be a significant change to the formulation or to the active pharmaceutical ingredient (API), or it may be a relatively minor change, such as extending the expiration date or changing a test method as technology improves. Whatever the modification, the key to any change is to understand the impact that the change will have on the safety and efficacy of the product. To do that, the right studies need to be designed to maximize the value of the data in order to demonstrate the impact – if any – of the change on the product. The design of these studies needs to take into consideration the type of change proposed, the type of regulatory notification required, and the technical data available regarding the product and the effect of

From: *Pharmaceutical Stability Testing to Support Global Markets*: Pharmasp,
Edited by: K. Huynh-Ba, DOI 10.1007/978-1-4419-0889-6_24,

the potential change on the product's stability (is the change already supported by previous development studies, is it within the design space, is information available that indicates the proposed change will have a significant impact?). All these points need to be considered in order to demonstrate the acceptability of the proposed change to the product to facilitate obtaining regulatory approval to implement it. In this paper, a review of FDA's categorization of product changes will be presented, followed by some case studies demonstrating how to maximize data for post approval changes.

2. Regulatory Reporting Categories

FDA defines three categories of change: major, moderate, and minor. Changes are classified based on the potential of the change to have an adverse effect on the product. With the increased potential to have an adverse effect, more time is required by FDA to review the change. A change that has the potential to have a substantial impact on the product is classified as a major change, which requires communication to FDA via a Prior Approval Supplement (PAS). FDA must provide approval before the changed product is distributed. Moderate changes are those that have the potential to have a moderate adverse effect on the product. These changes also require notification to the FDA in advance via Changes Being Effected (CBE) submission. The CBE-30 requires 30-day advance notice to allow FDA an opportunity to review the information. After 30 days, if no questions is received from the Agency, the changed product may be distributed. The CBE-0 submission allows for the distribution of the changed product upon receipt of the supplement by the FDA; there is no need to wait for any feedback. Those changes determined to have minimal potential to have an adverse effect on the drug product are classified as minor. These types of changes are communicated to the FDA in the Annual Report. The changed product may be distributed prior to notification of the change to the Agency, which will occur at the time of the next annual report. These different types of reporting categories and corresponding examples are described in the current guidance documents.[1]

Applicants have an option to potentially reduce the reporting category for the change by using a Comparability Protocol.[2] The Comparability Protocol is a comprehensive plan designed to demonstrate the lack of adverse effect on the product as related to the safety and efficacy of the product. The protocol must define the change(s) and the steps that will be taken to assess the impact on all the quality attributes potentially affected by that change. The protocol must describe the specific tests and studies that will be conducted, the acceptance criteria that must be met to determine that the changed product is comparable to the original product, and the reporting mechanism that will be used to communicate the change and supporting information to the FDA. The Comparability Protocol does require FDA's concurrence. It may be submitted as part of the original application or as a prior approval supplement. Upon FDA's concurrence, the plan may be executed.

---

[1] Guidance for Industry: Changes to an Approved NDA or ANDA, CDER April 2004.

[2] ICH Guideline: Guidance for Industry Q5E Comparability of Biotechnological/Biological Products Subject to Changes in their Manufacturing Process.

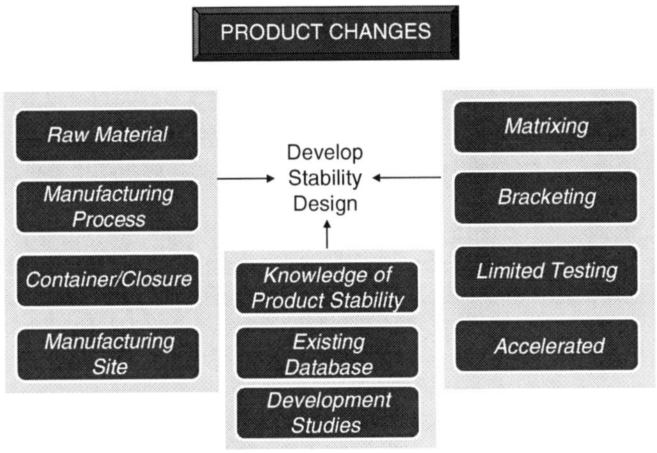

Fig. 24-1. Factors to consider when developing stability studies for product changes

3. Stability Study Designs

The key for any change, regardless of how it is to be reported, is to conduct the appropriate studies in order to assess the effect of the change on the product's identity, strength (e.g., assay, content uniformity), quality (e.g., physical, chemical and biological properties), purity (e.g., impurities and degradation products) or potency (e.g., biological activity, bioavailability, bioequivalence) as these factors may relate to the safety or effectiveness of the drug product. The studies must demonstrate that the change has no adverse effect on the product's stability. Figure 24-1 highlights factors to be considered when designing the appropriate stability studies for product changes. The proposed change needs to be understood – what exactly is changing and is there more than one change occurring? A review of the existing technical data for the current product is needed in order to understand the current product's stability profile and the factors that will or will not affect it. Is the proposed change within or outside the design space established by development studies conducted previously? What data are available regarding the change – for example, if the API is changing, how much new drug substance stability data are available? Does the profile look the same?

Potential study designs also need to be considered. If the change affects one product with only a few presentations, it may be appropriate to place a batch of each presentation on stability and generate comparative accelerated data to demonstrate any difference between the current product and the proposed product. If the change affects a product line consisting of multiple presentations, a matrixing or bracketing approach should be considered to demonstrate any impact of the change while minimizing the amount of testing required. It may also be possible to design studies with limited testing, with the intent to focus on characterizing those parameters that may be affected by the change.

Following are some case studies of product changes and the stability strategies developed to demonstrate the effect of the change (or lack of) on the finished product. In each case, the stability strategy and submission strategy were reviewed and agreed upon with the Agency prior to submission of the change.

3.1. Case Study 1

In the first case study, the source for the drug substance used in a parenteral solution was being changed to a new supplier. With the change to a new supplier, the material was also changing from the base form to the hydrochloride salt. Because of the change to the salt form of the API, a change to the manufacturing process for the finished product was also required. The pH adjustment step was eliminated as the drug was already in the salt form.

Upon review of the available drug substance stability data, it was determined that the new material had a different impurity profile; it contained a new peak, which was described in the supplier's Drug Master File (DMF).

This material is used in a frozen premix product, which has two presentations (same concentration, different sizes). The frozen product must be thawed prior to use; therefore, the product also has thawed condition label claims. For frozen products, the accelerated conditions are refrigerated (5°C) and room temperature (RT) conditions, which are also the thawed conditions. The stability trends under thawed conditions determine the product's expiry dating. The product has been in the market for over 15 years; therefore, an extensive stability database exists. The product's stability profile is well characterized. To demonstrate that there were no adverse affect of the changed API on this finished drug product, three primary stability batches were manufactured, two of the high dose, and one of low dose. The study design called for three batches to allow for a meaningful statistical analysis of the results. At the zero time test interval, a study was conducted at the thawed – or accelerated – conditions for 35 days at 5°C and 96 hours at RT, which went just beyond the product's thawed label claim. Under the accelerated conditions, the changed product demonstrated the same stability trends as the current product. The profiles were similar with the addition of the one new peak; the product still met all acceptance criteria.

For this major change, a Prior Approval Supplement was submitted with the initial release data, as well as the short-term data generated at the thawed or accelerated conditions. In addition, data from historical stability studies manufactured with the current supplier's API material were included for comparison purposes along with a commitment to conduct a safety study for the new peak and the typical stability commitments upon commercialization.

3.2. Case Study 2

The second case study is also an example of a change to active ingredients. In this case, four active ingredients were changing. The supplier was making changes to the manufacturing facilities, changing manufacturing sites, and for one of the active ingredients, a fermentation product, the starting strain was changing. One thing that didn't have any change was the manufacturing process for each active. A review of the supplier's API stability data indicated that the changed materials were comparable to the current materials.

These four APIs are used in a nutrition therapy line, which includes 10 product families with 129 different presentations. The products vary in formulations (the number of components present, the combination of components present, and the amount of each of the components present), in the size or volume of the different solution products, and in the container system (three different container systems are used across this product line). The products have been in the market for 10 years or more than that; again, an extensive stability database exists. The four specific active ingredients changing are very stable.

To demonstrate the effect of the changed APIs on the finished products, four stability batches were proposed, utilizing a bracketing approach to cover the 129 different presentations in the product line. The four batches representing four product families were selected to include the formulation with the highest concentration of the actives being changed (which would have the greatest potential for particulate formation), to bracket the container sizes (the smallest and largest container volumes were included), and to include the three different container systems used in the product line.

A Prior Approval Supplement with 3 months of room temperature and accelerated data from each of the four stability studies was submitted to each of the reviewing divisions. Upon approval of the PAS, as per the strategy agreed upon with the Agency in advance, CBE-0s were then submitted with the same data to each of the remaining applications in the product line, with a commitment to place the first production batch from each affected file on stability.

3.3. Case Study 3

The third example is a change to a new flexible container closure system for the primary container for a product line of terminally sterilized parenteral solutions. The new container closure system contained different plastic materials, and it required a change in the sterilization process for the solutions. The plastic materials being used in the new container system were qualified and had been previously approved for use with other solution products. The change in container system was being made to a product line containing solutions for fluid and electrolyte replenishment therapies. The product line consists of a variety of solutions containing a number of different electrolytes and simple sugars. The line includes 56 product types with a total of 91 presentations that vary by the levels of solutes present, which solutes are present and in what combinations, and in container size. All are stable solutions with well-characterized stability profiles.

With 91 product presentations, three batches per product would equal 273 studies, which would have a significant impact on a test laboratory. Therefore, the available technical information was reviewed and a reduced stability study design was developed that would demonstrate the acceptability of the new container system on the product line. The products were grouped into seven families based on the similarity of solutes present, the complexity of the formulation, the target pH of the product, and the ionic strength of the product. From this grouping of seven product families, four core products were selected for thorough evaluation; three batches (one of each container size) for each of these four core products were manufactured for 12 studies. These batches were also manufactured using containers fabricated from three unique plastic film lots. The four core products covered the range of all solute concentrations, the range of ionic strength and the range of target pH – all factors that could promote the most variation in interaction with the new plastic container system. An additional thirteen batches (one batch per product) were placed on stability to cover the range of minimum and maximum solute concentrations as well as the smallest and largest surface area to volume ratios, which is a critical consideration for flexible container systems.

Interval (mos)		0	1	2	3	4	6	9	12	15	18	24	36
Stability indicating	RT	X		X		X	X	X	X	X	X	X	X
	40C	X	X	X	X		X						
Weight Loss	RT	X		X		X	X	X	X	X	X	X	X
	40C	X	X	X	X		X						
Stable solutes	RT	X					X		X		X	X	X
	40C	X			X		X						

Fig. 24-2. Testing schedule with adjusted testing frequency for different parameters based on their impact to stability profile

For aqueous products packaged in semi-permeable container systems, water loss is a key stability parameter. Therefore, nine studies were added to the study design to monitor water loss across the thickness of the plastic sheeting used to fabricate the container system (three at the target thickness, as well as three at the upper tolerance and three at the lower tolerance).

With a study design that required 25 product stability studies and 9 water loss studies, the next step was to consider options to further reduce the testing load. For those parameters that were stability-indicating, which included the weight loss testing, testing was conducted at a greater frequency than recommended by the ICH guidance[3]; testing was scheduled at the 2- and 4-month intervals instead of a 3-month interval, and a 15 month interval was added to the schedule. For those solutes that are very stable and would not be affected by the container system, testing was scheduled every 6 months. Figure 24-2 illustrates the testing schedule followed.

To support this change to 56 products with 91 presentations, 25 solution stability studies and 9 weight loss studies were conducted with some reduced testing. An update was submitted to the DMF (Drug Master File) for the container system. For the solution product applications, PAS submissions were prepared with 6 months of accelerated and room temperature data and a commitment to place the first commercial batch of each product onto stability to cover all of the product families included in the container system.

4. Summary

The key to the successful study design is to understand what is changing, what the regulatory reporting requirements are, and what technical information is available regarding the product and the potential impact of the change. Understanding the product and the proposed change allows for the design of

[3] ICH Guideline: *Guidance for Industry* Q1A(R2) Stability Testing of New Drug Substances and Products.

appropriate stability studies that are also the most efficient and maximize the data to demonstrate the acceptability of the change to gain regulatory authority concurrence. For each of the case studies, the proposed strategy was submitted to and agreed upon with the Agency in advance. The stability studies were conducted and successfully demonstrated the acceptability of the proposed changes on the products. The study designs demonstrated no impact to product safety or efficacy, which resulted in FDA's approval of the product changes.

Author Biography

Mrs. Paula Youngberg Webb is Senior Director of the Stability Operations, Global R&D group at Baxter Pharmaceutical Technologies, Baxter Healthcare Corporation. She is responsible for management of the US stability programs for various divisions within Baxter Healthcare Corporation. The products developed by these divisions include intravenous premix drug solutions, parenteral nutritional solutions, renal dialysis solutions and products supporting blood therapies. These responsibilities include products under development as well as commercial products. In addition, Paula provides R&D support for FDA inspections. After receiving an M.S. degree in Analytical Chemistry, from Wayne State University, Paula joined Baxter as a chemist in the method development laboratory. From there, she had roles in the stability organization followed by managing the Technical Quality unit, which supported the Scientific Affairs R&D organization. She rejoined Stability Operations in her current role in June 2004. She's currently a member of American Association of Pharmaceutical Scientists, American Society of Quality and PDA.

Chapter 25

Use of Statistics to Establish a Stability Trend: Matrixing

Earl Nordbrock

Abstract Guidance Q1E gives several options for calculating shelf life. These options are compared using examples. Bracketing and matrixing (Guidance Q1D) are the ways to reduce the amount of testing in a stability study. Examples of matrixed and bracketed designs are presented, and the probabilities of obtaining a given shelf life for the reduced designs are compared to the same probabilities for the full design.

1. Introduction

ICH Q1E presents several methods to analyze long term stability data. When there are one-strength of product, one package, and three batches, then Q1E has three methods for analyzing the data. The first example compares these three methods. Data are the assayed strength of a product with a specification 90–110% of label claim (LC). It is assumed that potency can not increase, so the 95% lower confidence bound is required to be greater than 90% LC at all times prior to the calculated shelf life. In method 1, each batch is analyzed separately; in method 2 all three batches are analyzed with one analysis using individual intercepts, individual slopes and pooled mean squared error; and in method 3 all the three batches are analyzed with one analysis and the poolability of batch slopes and batch intercepts are tested.

For method 1 each batch is analyzed separately, and a separate regression line and a separate 95% lower confidence bound are calculated for each batch. Data from the first batch are analyzed in Fig. 25-1, and the calculated shelf life is 22 months. Data from the second and third batches are analyzed in Figs. 25-2 and 25-3 respectively, and the calculated shelf lives are 21 months and 16 months. The calculated shelf life of the product is the minimum of these three numbers, or 16 months.

The same data are analyzed using method 2 in Fig. 25-4. There is one analysis where each batch has a separate intercept and a separate slope. The regression line for each batch is the same as that found in method 1, but the confidence bounds are different as the mean squared errors from the three batches are pooled before calculating confidence bounds. Because the mean squared

From: *Pharmaceutical Stability Testing to Support Global Markets*: Pharmasp,
Edited by: K. Huynh-Ba, DOI 10.1007/978-1-4419-0889-6_25,

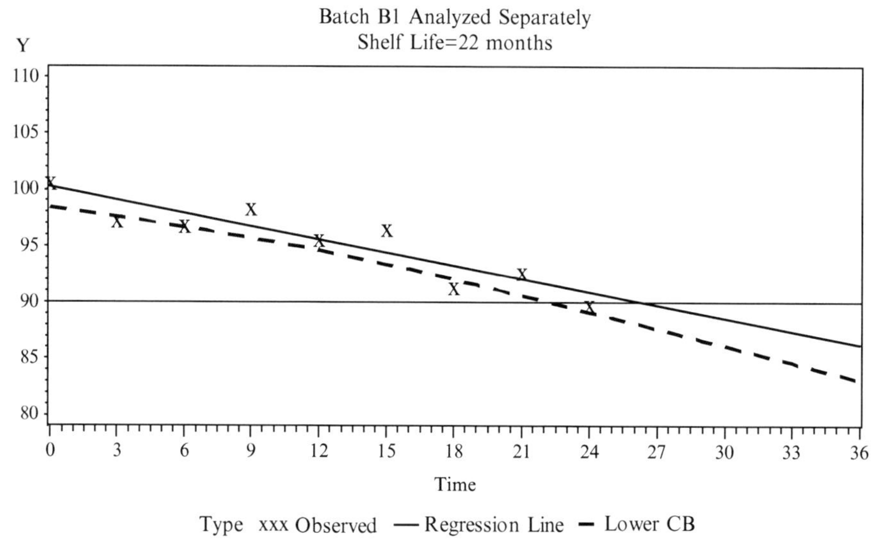

Fig. 25-1. Analyze Batch B1

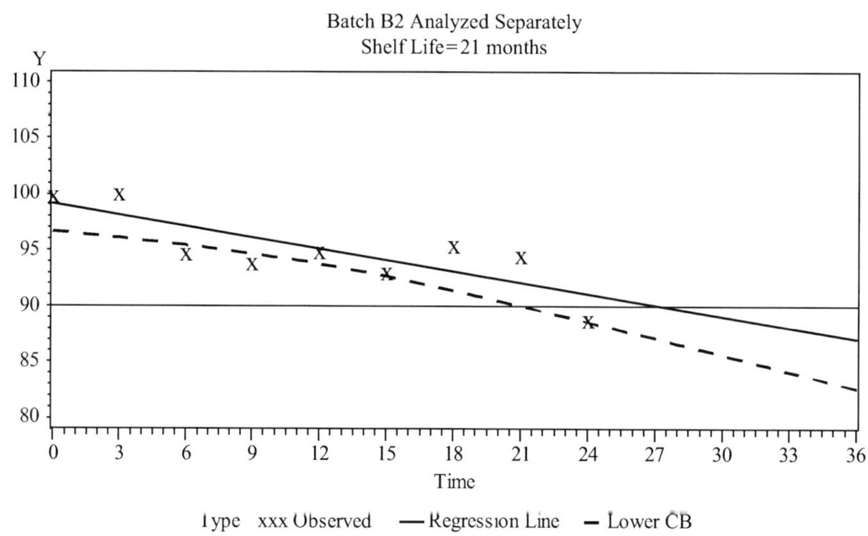

Fig. 25-2. Analyze Batch B2

errors from the three batches are pooled, method 2 typically has confidence bounds that are closer to the regression lines than are those in method 1, and thus typically results in a longer calculated shelf life than does method 1. The calculated shelf lives for the three batches are 21, 21, and 17 months. Thus the calculated shelf life of the product is 17 months.

The same data are analyzed using method 3 in Fig. 25-5. There is one analysis, but this time the poolability of batch slopes and the poolability of batch intercepts are tested. The p-value from the test for poolability of batch slopes is 0.43, and that for poolability of batch intercepts is 0.41. Since both are greater than 0.25 the final reduced model is a common slope common intercept model. The calculated shelf life is 21 months.

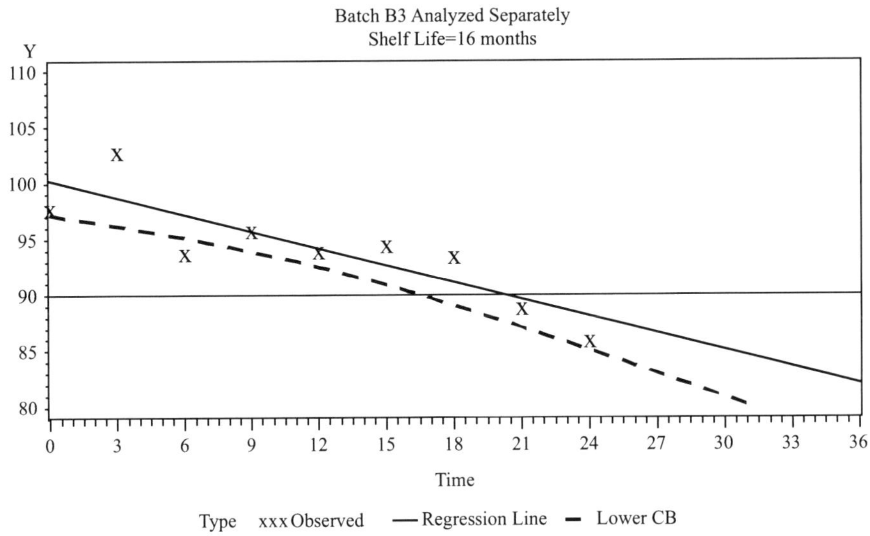

Fig. 25-3. Analyze Batch B3

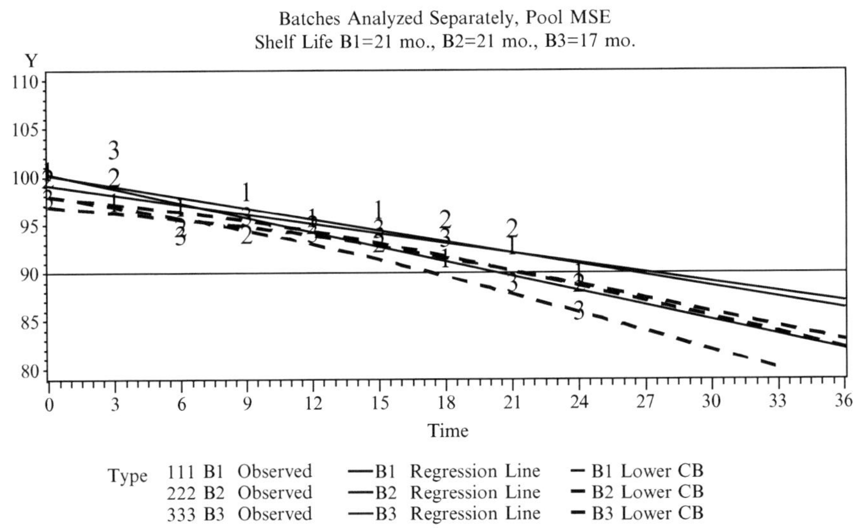

Fig. 25-4. Batches Analyzed Separately

In summary, the calculated shelf life of the product is 16 months using method 1, 17 months using method 2, and 21 months using method 3. Typically the calculated shelf life will be longer with method 3 than method 2, and longer for method 2 than method 1. However all three methods are acceptable according to Q1E.

From all three analyzes, it is clear that it is common to have observations which are less than the 95% lower confidence bound for the mean. This points out a conflict between Q1E and the OOS guideline (Investigating Out-of-Specification (OOS) Test Results for Pharmaceutical Production, October 2006). The OOS guidance requires an investigation for every observation that is OOS, and thus in effect the OOS guidance requires every observation to be within specification. However, Q1E states that "the mean value...can be expected to remain within the acceptance criteria", and there is no requirement

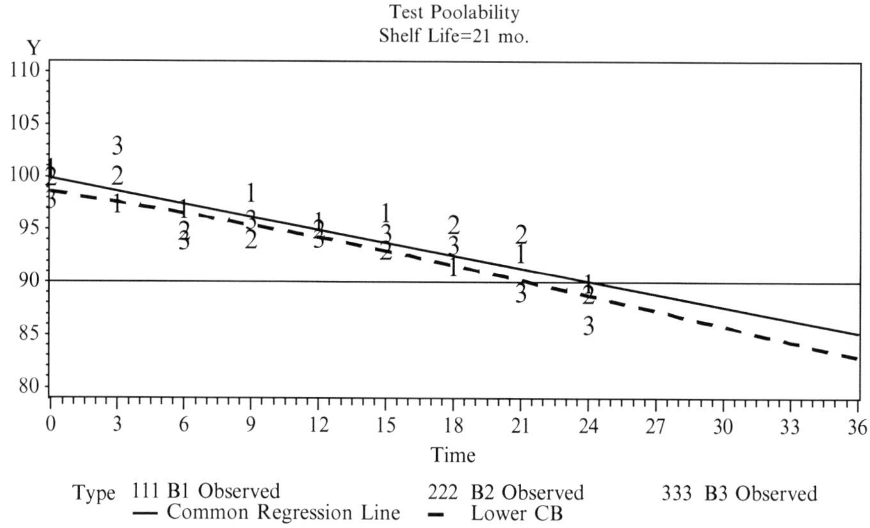

Fig. 25-5. Batches Analyzed by Testing Poolability

that all individual values need to be within the specification. In fact, up to 50% of individual observations can be less than the 95% lower confidence bound for the mean at the calculated shelf life.

2. Statistical Designs

Let us now turn to stability study design. The design which gives the best estimate of the slope has one-half of the observations at the start of the study (tzero) and one-half at the end of the study, say 36 months. This design minimizes the variability of the slope and has the shortest confidence interval for the slope of any design. On the other hand, the best design to determine whether mean potency is within specifications at say 36 months is to take all observations at 36 months. Obviously neither of these designs is practical for a stability study, but the concepts can be used to design stability studies. These two concepts indicate that a good statistical design has less testing in the middle of a study than at the beginning or the end. A matrix design is a stability study which has only a portion of all samples tested at specified time points. The time points between the start and the end are not as important, and thus a good matrix design has the number of tests reduced at these intermediate times. However, because a submission is often made with 12 months data, then testing at 12 months is important, and the 12 month testing is not reduced.

Table 25-1 is a long term matrixed study design, with testing reduced to one-third at 3, 6, 9, 12, and 24 months, while retaining full testing at 0, 18, and 36 months. It is assumed that analyses will be done after 18 and 36 months. The risk when doing this matrixed design is minimal. For example, if the RSD is 2.0%, population slope is 2.0% per year, specifications are 90–110%, and the population release value is 100%, then at 18 months the full design has a 69% chance of obtaining a calculated shelf life greater than 30 months, whereas the matrixed design has a 63% chance. Thus the difference between the matrixed design and the full design is minimal.

Table 25-1. Long term matrix (1/3 to 3/3) design.

Batch	Times tested (months)
Batch 1	0, 3, , , 12, 18, , 36
Batch 2	0, , 6, , , 18, 24, 36
Batch 3	0, , , 9, , 18, , 36

Table 25-2. Unacceptable long term matrix design.

Batch	Times tested (months)
Batch 1	0, 3, 6, 9, 12, 18, 24, 36
Batch 2	0, , , , , 18, , 36
Batch 3	0, , , , , 18, , 36

Table 25-3. Upright inverted study.

Batch	Upright	Inverted
Batch 1	T1	T2
Batch 2	T2	T1
Batch 3	T1	T2

T1 >test at 0, 3, , 9, , 18, 24, 36
T2 >test at 0, , 6, , 12, 18, 24, 36

It is wise to test time 0 samples in triplicate. Triplicate ideally means three different samples done on three different days by three different analysts using three different machines, as it is desired to obtain the same variability that will be observed later in the study. The rational behind triplicate initials is that these initials are very important, as they often apply to all storage conditions and all packages. For example, initials are used to evaluate whether there is a significant change at the accelerated condition and the same initials are used to calculate the shelf life of the long term storage condition. Thus if the 6 months results appear questionable one can always repeat the analysis, maybe at 6.5 months. But if the initial looks out of trend after 6 months, there is no way to obtain another initial. Thus it is wise to do triplicate initials.

An unacceptable long term matrix design is presented in Table 25-2. This design is unacceptable because batches aren't tested similarly. Namely, Batch 1 is tested very often, whereas Batches 2 and 3 are tested only when an analysis is going to be done.

Table 25-3 contains a matrix design for a study where the package must be tested both upright and inverted. Here every batch is tested at each time point, but testing alternates between upright and inverted within each batch. Also, at every time either one or two of the batches are tested upright and the remaining batches are tested inverted.

Consider a study where there are three batches of product, and where each batch of product is packaged into three different HDPE bottles. The full design would be to test each package at each time point. A matrix design for this study is presented in Table 25-4. Using the same population parameters used previously, at 18 months the chance of obtaining a 30 month shelf life is 65%

Table 25-4. Matrix design-product with three packages.

Batch	Package		
Batch	**HDPE30**	**HDPE100**	**HDPE200**
B1	T1	T2	T3
B2	T2	T3	T1
B3	T3	T1	T2

T1>test at 3, , , 12, 18, , 36
T2>test at , 6, , , 18, 24, 36
T3>test at , , 9, , 18, , 36

Table 25-5. Unacceptable matrix design-product with three packages.

Batch	Package		
Batch	**HDPE30**	**HDPE100**	**HDPE200**
B1	T1	T2	T3
B2	T1	T2	T3
B3	T1	T2	T3

T1>test at 3, , , 12, 18, , 36
T2>test at , 6, , , 18, 24, 36
T3>test at , , 9, , 18, , 36

Table 25-6. Bracket design-product with three packages.

Batch	Package		
Batch	**HDPE30**	**HDPE100**	**HDPE200**
B1	T		T
B2	T		T
B3	T		T

T>test at 3, 6, 9, 12, 18, 24, 36

for the matrixed design, whereas for the full design it is 67%. Again, the risk of using the matrixed design is minimal.

An unacceptable matrix design for this product with three packages is presented in Table 25-5. This design is unacceptable because the testing among the packages is not balanced. For example, only one package is tested at 3 months, whereas in a balanced design every package would be tested at 3 months.

Another design which reduces the amount of testing is a bracket design. In a bracket design, only the extremes are tested. The extremes may be the smallest and largest HDPE, the smallest and largest strength if all strengths are made from the same exact formulation, etc. The choice of the extremes must be defended, as the extremes may be the amount of head space and not the number of tablets in a bottle. A bracket design for a product with three packages is presented in Table 25-6. In this example, only the smallest and largest bottles are tested.

One naturally asks whether matrixing and bracketing can both be applied in the same study, and the answer is yes. The amount of testing in the bracket design in Table 25-6 can be reduced by matrixing, as given in Table 25-7.

Table 25-7. Matrix and bracket design-product with three packages.

Batch	Package		
Batch	**HDPE30**	**HDPE100**	**HDPE200**
B1	T1		T2
B2	T2		T1
B3	T1		T2

T1>test at 3, , 9, , 18, 24, 36
T2>test at , 6, , 12, 18, , 36

3. Analysis of Matrixed and Bracketed Designs

The analysis for matrixed and bracketed designs is the same as the analysis of full designs. Although multiple methods of analysis are presented in Q1E, the longest calculated shelf lives are typically found when testing the poolability of slopes and intercepts across batches, across packages, and across strengths. Terms related to batch are tested at the 0.25 level of significance, and terms not related to batch (e.g., strength and package) are tested at the 0.05 level of significance. In a bracketed design, the missing shelf lives are assigned the shorter of the shelf life of the extremes.

4. Prior Approval

When doing a matrixed or bracketed design for US submission, one should get prior approval from the FDA. The FDA often lets the sponsor update the stability analysis during the review process, e.g., when the 18 month analysis is available, thus reduced testing is done at the 12 month time but full testing is done at 18 months. However, the sponsor must obtain prior approval to use this strategy. In Europe, prior approval of the design is not required.

Author Biography

Dr. Earl Nordbrock is a self employed statistical consultant for the pharmaceutical industry. After receiving his Ph.D. in statistics from the University of Wisconsin, he taught statistics at the University of Alberta. He then joined the pharmaceutical industry, and has held various positions at McNeil, Wyeth, Marion Merrill Dow, Anesta, Pfizer and Purdue Pharma. He has more than 20 years experience in non-clinical statistics and is a recognized expert in the design and analysis of stability studies, including matrixed and bracketed studies.

Chapter 26

Setting Specifications for Drug Substances

Jon V. Beaman

Abstract This article discusses the purpose of a drug substance specification and the variety of tests to be considered for inclusion. The relevance of these aspects to stability studies is examined and the differences between release testing and tests relevant to stability studies are explored. A scientific approach to setting drug substance specifications and to designing stability studies is encouraged and where necessary, the regulatory landscape is compared and contrasted to this approach.

1. Introduction

The three important functions of an Active Pharmaceutical Ingredient (API) specification are: (1) to ensure patient safety regarding any potential API characteristics that are not to be controlled via the drug product specification (e.g. process related impurity levels); (2) to ensure that the API is of suitable quality such that a drug product can be successfully manufactured; (3) to ensure the API batch characteristics will not negatively impact the ability of the drug product to meet its specification throughout its shelf life. It should be noted that the latter two functions of the API specification are in fact business driven, i.e. if these are not in place on the API specification, then the impact will be a subsequent failure of the drug product to meet its specification (e.g. potency or dissolution requirements) rather than there being any impact upon patients. This should be borne in mind when considering what should be included in a (regulatory) specification as we move forward into a 'Quality by Design world' (see discussion in last section).

In addition to determining the tests and limits that need to be included in the API specification, the development scientist will, in parallel, need to determine which characteristics should be measured during stability studies. Many items that are included in an API specification are not "stability indicating", that is they are not parameters that need to be measured over time due to the fact

[1] ICH Guideline Q1A(R2) Stability Testing of New Drug Substances and Products (Feb. 2003).

From: *Pharmaceutical Stability Testing to Support Global Markets*: Pharmasp,
Edited by: K. Huynh-Ba, DOI 10.1007/978-1-4419-0889-6_26,

that they are fixed at the time of manufacture (e.g. heavy metals) or perhaps have been shown not to change over time during earlier forced degradation or stability studies. International Conference of Harmonization (ICH) guideline Q1A(R2) states that stability evaluation should cover those features that are susceptible to change during storage and likely to influence quality, safety and/or efficacy.[1] It may also be that the parameters additional to those included in the specification are measured during stability studies in order to build an understanding regarding the behaviour of the API on storage.

Note that, although ICH guideline Q6A states "The concept of different acceptance criteria for release vs. shelf-life specifications applies to drug products only" there may be instances where tighter limits need to be applied internally to the API at the time of release e.g. for degradation products.[2]

2. Developing Stability Testing Strategies Appropriate to the Stage of Development

Most guidelines were written to provide guidance on registration requirements. The ICH guidelines in particular were not intended to be applied to investigational materials. Within the ICH guidelines, it is stated "This guideline does not apply to potential new drug substances…used during the clinical research stages of development". During early stages of development, the process and product understanding is likely to be much less than that possible at the registration phase. By the time of registration, the years of building knowledge through experimental work and the manufacture of a number of batches means that the developer has built up a comprehensive understanding of the API, its pharmaceutical performance and therefore the parameters to be controlled and the limits appropriate. Table 26-1 outlines possible approaches to control at various stages of development.

Regarding the allocation of shelf lives to APIs, the US and EU guidelines do not actually specify that the results from stability studies are required to

Table 26-1. Outline approach to API control at various stages of development.

Stage of development	Approach
Pre-first in human	Internal control only; generic methodologies; identity test and report values for potency & purity. Assigned retest date based on purposeful degradation and/or accelerated testing and chemistry knowledge
Exploratory phase (approx. phase I/IIa)	Developing knowledge; key safety (e.g. impurity) tests on specification with defined limits, however information may be gathered internally on other parameters on release and/or stability to build a fuller picture; acceptance criteria based more on scientific rationale rather than on (limited) batch data. Note: FDA guidance states: "…established specifications ordinarily need not be submitted at the initial stage of drug development"[a]Exposure to drug is more highly controlled in early clinical phase; controlled environment; small number of patients; limited duration of dosing
Development phase (approx. phase IIb/III)	Generally more aligned with the ICH guidelines/other registration requirements; however relatively limited batch data at this stage and/or specifications may be being developed in line with QbD principles (see discussion in last section)

[a] FDA content and format of investigational new drug applications (INDs) for phase 1 studies of drugs 11/1995

[2] ICH Guideline Q6A Specifications: Test Procedures and Acceptance Criteria for New Drug substances and New Drug Products (Oct. 1999).

be submitted in INDs or IMPDs, and essentially the requirements are for a description of the stability studies to support the trial, plus any available data as it becomes available.[3,4] While the expectation is that some measure of stability data is available at time of IND submission to demonstrate the stability of drug substance and drug product through the retest period, the only true requirement is that the stability studies on the clinical materials are conducted concurrent with the length of the clinical trial so that sponsors can assure regulators that the materials are stable through the investigational period.

3. Specific Tests to be Considered for APIs

When developing specifications and stability testing strategies, the developer should consider the target product profile and the functions of a specification (as listed) before including requirements and also before setting limits. In addition, for stability studies it is unnecessary to include testing criteria that are not both: (1) 'stability indicating' i.e. have the potential to change on stability, and (2) have a potential to compromise patient safety or drug product manufacture or performance.

Care should also be exercised when using batch data to determine specification limits, otherwise artificial and unreasonable limits may be applied to parameters based on limited batch data even though higher limits may not compromise either patient safety or drug product manufacture and performance i.e. a scientific and safety based approach to setting specifications and limits should be encouraged. This approach is discussed in more detail in the Impurities section as well as the last section of this paper.

In the following sections, some of the more common tests are discussed and recommendations are made as to when the tests should be considered for specification and stability testing purposes, particularly from a scientific and product relevance point of view.

3.1. Appearance

Acceptance criteria are frequently "white to off-white powder"; in Japan "JP white" covers this range. Colour descriptions are available, including a recommendation of "white to almost white".[5] Note that appearance is only a qualitative test and genuine safety related concerns (e.g. coloured impurities present at a significant level) should be dealt with via appropriate quantitative tests.

3.2. Solution Clarity

This test is sometimes performed during development for APIs intended for solution formulations and is generally performed for information rather than being part of the API specification. Particulates, if important as regards

---

[3]EMEA Requirements to the Chemical and Pharmaceutical Quality documentation concerning investigational medicinal products in clinical trials. CHMP/QWP/185401/2004final March 2006. http://www.emea.europa.eu/pdfs/human/qwp/18540104en.pdf

[4]FDA Guidance for Industry: INDs for Phase 2 and Phase 3 Studies Chemistry, Manufacturing, and Controls Information. http://www.fda.gov/cder/guidance/3619fnl.htm

[5]Technical Guide for the Elaboration of Monographs 4th Edition 2005 European Pharmacopoeia.

product safety, should be tested as part of the drug product specification. Any trends below normal visible observation can be followed using fibre optic light, e.g. during stability studies; this testing may give a heads-up as to future potential visible observations at future checkpoints.

3.3. Turbidity

Turbidity is a test that can provide a more objective numerical value to visible solution clarity observations. Generally results can be expected to be below around 2 NTU. In an attempt to determine a link with visible observations, we compared the EP test solution for ‘clear’ from the EP Clarity and Degree of Opalescence test (actually a drug product test) and found a reading of 2.8 NTU corresponded to a just-visible observation. Note however that turbidity results depend on a number of factors including refractive index and particle size of species, and therefore, the ‘visible threshold’ can vary. The test can be useful however for monitoring trends on stability and providing numerical values.

3.4. Identity

Identity is a test required at release but is an unnecessary test during stability studies. During release testing, chiral identity may also be required if it is feasible that the enantiomer may be formed during manufacture instead of the intended chiral form. If using a spectroscopic technique to demonstrate identity, then a simple optical rotation test is adequate to demonstrate correct chirality. Spectroscopic tests for the API may be specific for the salt form; however if not, a specific test for the counterion identity may be required for release testing.

3.5. Solid form Identity

If a number of forms are known to exist then a test for solid form identity at release may be required. However, unless forms other than the desired form are feasibly formed during the API manufacturing process, a test would not be scientifically justified. If other forms were feasible, the development scientist would then consider whether the solid form affected relevant dosage form parameters before determining whether a test was required. In addition, unless development work demonstrated that other forms may be formed at relevant temperatures/humidities or timescales (e.g. hydrates, amorphous crystallisation), a test should not be required during stability studies. It should be noted that in addition to potentially different stability characteristics for different polymorphs, significant changes to crystal habit should be investigated during development for any effects on stability (Hancock and Zografi 1997). Again a test should not be required on stability.

If batch to batch variability in degradation rates is noted then it may be prudent to examine levels of amorphous material in the API batches. This is because reactivity (including degradation) in the amorphous state is greater than that in the crystalline state (Carstensen and Morris 1993). If this was found to cause significantly different degradation rates between batches whereby the shelf life was affected, a test at release may be required; sensitivities down to around 5% may be achievable with PXRD (Hancock and Zografi 1997) and 1% by microcalorimetry (Humera et al. 1996).

3.6. Particle Size

Particle size may affect formulation processing and stability characteristics and may form part of the release testing strategy. However, it should not be necessary to monitor particle size on stability as either particle size reduction or growth are unlikely in the solid state. Also, it should be noted that usually the particle size is only relevant in that it may affect drug product manufacture or performance, which in turn should be measured directly as part of the drug product specification (e.g. potency or dissolution). PQRI recommendations for methodology and acceptance criteria are available, which state that particle size "may be performed as an in-process test or…release test." (Snorek et al. 2007). Probably, the only potential concern on stability is agglomeration of particles. If either powder processing issues are noted during formulation manufacture or the performance of the dosage form is affected by the use of API stored for longer periods, agglomeration of API on storage may be occurring. If confirmed (e.g. via microscopy (Nichols et al. 2002)) this is likely to be resolved by a change of storage conditions or the use of dessicant for the API, a reduction in shelf life of the API or a change to the formulation manufacturing process, rather than a long-term need to include particle size testing on stability studies.

3.7. Assay

This requirement is included for release and stability testing. A frequently adopted upper limit is 102.0% (on an anhydrous or dried i.e. anhydrous and solvent free basis); the lower limit usually depends on the potential levels of impurities (Carstensen and Morris 1993). For racemic drugs with enantiomers of essentially equal efficacy/safety, then the assay requirement can be as a total of the two. It can be argued that assay testing during API stability studies adds little value over doing purity testing alone and a justification for omitting an assay test on stability can be considered.

During the later stages of development, assay determination of the counterion could be omitted from the release specification if batch data shows good stoichiometry routinely achieved. Note that, usually, the counterion would only be monitored on stability if it is known to potentially degrade.

3.8. Water

Water level in an API is in itself not a patient safety risk and may only indirectly effect the other performance characteristics of the API or the formulation. During earlier stages of development, it is unnecessary to constrict water limits by specification; water is really measured for information purposes to link to important parameters, which will be measured directly and controlled, such as degradation product formation. The requirement for a specification test should be determined based on the understanding of degradation mechanisms of the API, the packaging to be employed and stability knowledge gained during development. If it is found that water content has an important effect on stability or formulation characteristics, then a test may be included on the specification.

[Ref: Skrdla PJ et al. (2009) Use of a Quality-by-Design approach to justify removal of the HPLC weight % assay from routine API stability testing protocols, J. Pharm and Biomed Analysis, *Available online 21 June 2009*]

3.9. Impurities

Process related impurities, which are not also potential degradation products, would be controlled during release testing only. ICH guideline Q3A (R2) outlines impurity control requirements at the registration phase.[6] It states that where there is no safety concern, impurity acceptance criteria should be based on the data generated on batches of the new drug substance manufactured by the proposed commercial process, allowing sufficient latitude to deal with normal manufacturing and analytical variation and the stability characteristics of the new drug substance. However, this thinking does not fit well with respect to the principles of Quality by Design. The ICH Q3A (R2) guideline leads to limits essentially being based around limited batch experience and unnecessarily constricts future process development, even when there are no patient safety concerns.

Similarly, limits for degradation products should not be bounded within actual data available at the time of filing; ICH Q6A states: "estimate maximum increase in impurity at retest date". Thus it should be feasible to extrapolate degradation product levels, explaining the basis for doing so in the specification rationale, to a point beyond the initial interim retest date, providing the stability study is due to continue until this timepoint and providing the degradation product does not become a safety concern at the levels predicted. Note that, when developing limits for degradation products, if the degradation product(s) are common to both the API and the drug product, the API specification limits should be constricted to ensure a workable shelf life for the drug product.

If it is feasible that the opposite enantiomer could be formed during the manufacture of a chiral API, then a test for the enantiomer may be required at time of release. However if development work (scientific analysis, stress testing) shows that the opposite enantiomer is not a degradation product, then a test on stability is not scientifically required.[7]

An additional impurity control requirement is for total impurities. However, the assay and individual impurity controls ensure efficacy and safety and therefore a total impurity requirement does nothing to further assure patient safety or batch efficacy. Perhaps, this requirement might be reviewed at some point in the future…

3.10. Solvents

Solvents are required to be controlled appropriately on the API specification according to ICH guideline Q3C(R3).[8] However unless one of the degradation products is also a solvent, then testing is not required during stability studies. Solvent levels may occasionally be determined as part of an assay/potency procedure if mass balance may be impacted by loss of solvent on stability.

3.11. Inorganic Impurities

Inorganic impurities may be controlled as part of release testing but would not be affected at the temperatures used for pharmaceutical stability testing and therefore it is not necessary to be tested as part of the stability study. Some specific considerations to be included on the specification include:

---

[6] ICH Guideline Q3A(R2) Impurities in New Drug Substances (Oct. 2006).
[7] FDA Development of New Stereoisomeric Drugs. 5/1/92 Last update: July 6, 2005. http://www.fda.gov/cder/guidance/stereo.htm
[8] Guideline Q3C(R3). Impurities: Guideline for Residual Solvents (Nov. 2005).

Heavy metals: this is essentially an unnecessary test if heavy metals are not used in the manufacture of the API and starting materials and reagents are adequately understood or controlled. Suggested criteria and limits are contained in the the EP Technical Guide (Carstensen and Morris 1993). For control of catalysts used during manufacturing process the EMEA guideline should be consulted.[9] Arsenic control may be required specifically for Japan.[10]

Residue on ignition: in reality, this is only a quality test. Generally accepted limits are 0.1% (Carstensen and Morris 1993),[11] although for Japan 0.10% may be requested.

Other: ICP-MS scans may be performed during the development to gather information. Specific tests may need to be developed depending on formulation/ other considerations, e.g. iron control due to possible iron oxide/hydroxide formation and precipitation in alkaline solution formulations.

3.12. Microbes and Endotoxins

Scientifically speaking, endotoxins should be a release test only and only if the API is intended for a parenteral drug product. Total viable aerobic count may be required if the API is for a sterile drug product, but may be omitted from specification based on ICH Q6A Decision Tree #6 (process steps, capability of supporting growth) plus water activity considerations. If tested on stability, consideration should be given to performing these tests at key checkpoints only, e.g. annually. Similar considerations apply to specific organisms, but this is essentially a quality/contamination test and is not relevant with regards to stability testing.

3.13. Other Tests to be Considered

Taste testing is not appropriate due to safety considerations and similarly odour is generally not performed (Carstensen and Morris 1993). Occasionally, odour may need to be considered depending on route of delivery and developers may need to limit solvents to levels tighter than those allowed in ICH Q3C (R3) if unpleasant odours are noted. Degradation products may also be odorous e.g. sulphates or toluene, and may need to be controlled to tighter limits than the safety considerations alone would allow.

4. Stability Commitment Batches

Although the stability study design for commitment batches may initially be based on the registration stability study design, innovators should consider scientifically justifying the removal of tests or checkpoints for commitment stability studies if information gathered during the registration stability studies justifies this.

---

[9] EMEA Guideline on the Specification Limit for Residues of Metal Catalysts or Metal Reagents. EMEA/CHMP/SWP/4446/2000 (Feb. 2008). http://www.emea.europa.eu/pdfs/human/swp/444600enfin.pdf

[10] Japanese Pharmacopoeia 15th Edition (Apr. 2006).

[11] FDA Drug Substance: Chemistry, Manufacturing, and Controls Information (2004 Draft.) Withdrawn as per FR notice June 1, 2006. http://www.fda.gov/cvm/Guidance/guide169.pdf

5. Stability Studies Final Discussion

A large proportion of (small molecule) APIs can be defined as stable when using the definition contained within Annex 1 of the EMEA guideline on stability testing for variations to a marketing authorisation.[12] The author reviewed a number of recent submissions and found that anywhere between 2 and 40 degradation products were seen for each compound during forced degradation experiments…but all of the APIs (and a majority of the solid products) showed no significant degradation on real time stability. The latest WHO stability guidance recognises that waiting for 12 months stability on three registration stability batches of API is frequently unnecessary and states that 6 months data may be submitted for APIs that are known to be stable.[13]

Another aspect of API stability that seems to have been confused with the need to demonstrate manufacturing capabilities is the term ‘representative’ when applied to stability batches. For stability study relevance, representative should mean parameters such as polymorph, water content and particle size – and even the relevance of these to stability performance depend on the particular API. Thus, even though starting materials, route or process to the API may change (and earlier batches may therefore not have been manufactured according to the commercial route or process) there may be no reason to suspect that the stability characteristics of earlier batches will not be relevant to the commercial API. It is likely that too many scientifically unnecessary API stability studies are triggered by default due to changes such as the route or process, which may be irrelevant to stability performance. Similarly the size of the batch is completely irrelevant to understanding stability, and the key characteristics of an API that affect degradation; it is only when scale-up affects key parameters that influence stability that an evaluation should be performed as to whether further stability studies are necessary. In this vein commitment, usually stability requirements are scientifically unnecessary as would be any annual/ongoing stability studies for APIs.

As we embrace continuous processing, where manufacturing parameters may be continually modified based on analytical feedback from the output, the conventional concept of a batch will be irrelevant (e.g. output may not be homogeneous although adhering to the specification) and the ‘stability design space’ of an API will need to be understood instead. Defining the stability space (i.e. understanding which parameters affect the stability, to what extent and where edges of failure to achieve a set shelf life are) will mean scientific understanding of the stability characteristics of the API, perhaps gained through purposeful accelerated stability experiments (Waterman et al. 2007) and in-silico modelling rather than empirical “3 batches, 12 months” stability.

---

[12]EMEA CHMP Guideline on stability testing for applications for variations to a marketing authorisation CPMP/QWP/576/96 Rev.1 (2005) http://www.emea.europa.eu/pdfs/human/qwp/057696en.pdf

[13] Stability testing of active pharmaceutical ingredients and finished pharmaceutical products. In: WHO Expert Committee on Specifications for Pharmaceutical Preparations. Forty-third report. Geneva, World Health Organization, 2009, Annex 2 (WHO Technical Report Series, No. 953). <http://www.who.int/medicines/publications/pharmprep/pdf_trs953.pdf#page=101>

6. Relevant Guidelines not Referenced Directly

EU Guidelines

- Chemistry of New Active Substances CPMP/QWP/130/96, Rev.1. (Feb 2004). http://www.emea.europa.eu/pdfs/human/qwp/013096en.pdf
- Annexes to ICH Residual Solvents Guideline. CPMP/QWP/450/03 (Feb 2005). http://www.emea.europa.eu/pdfs/human/qwp/045003en.pdf
- Guideline on the limits of genotoxic impurities. EMEA/CHMP/QWP/251344/2006 (Jun 2006). http://www.emea.europa.eu/pdfs/human/swp/519902en.pdf
- European Pharmacopoeia.

US Guidelines

- Guidance for Industry, Investigators, and Reviewers Exploratory IND Studies 1/2006 http://www.fda.gov/cder/guidance/7086fnl. htm
- Development of New Stereoisomeric Drugs. 5/1/92 Last update: July 6, 2005. http://www.fda.gov/cder/guidance/stereo.htm
- Drug Substance: Chemistry, Manufacturing, and Controls Information 1/2004 Draft. Withdrawn as per FR notice June 1, 2006. http://www.fda.gov/cvm/Guidance/guide169.pdf
- Stability Testing of Drug Substances and Drug Products 6/5/1998 Draft. Withdrawn as per FR notice June 1, 2006.
- BACPAC I: Intermediates in Drug Substance Synthesis; Bulk Actives Post-approval Changes: Chemistry, Manufacturing, and Controls Documentation 2/2001. http://www.fda.gov/cder/Guidance/3629fnl.htm Withdrawn as per FR notice June 1, 2006.
- United States Pharmacopeia.

International Guidelines

- South Africa Medicines Control Council "Stability" June 2006 v3.

References

Carstensen T, Morris T (1993) Chemical stability of indomethacin in the solid amorphous and molten states. J Pharm Sci 82:657–659

Hancock BC, Zografi G (1997) Characteristics and significance of the amorphous state in pharmaceutical systems. J Pharm Sci 86(1):1–12

Humera A, Buckton G, Rawlins DA (1996) The use of isothermal microcalorimetry in the study of small degrees of amorphous content of a hydrophobic powder. Int J Pharm 130:195–201

Snorek SM, Bauer JF, Chidambaram N, Doub WH, Duffy EP, Etzler FM, Kelly RN, Lane JJ, Mueller RL, Prasanna HR, Pujara CP, Reif VD, Scarlett B, Stowell JG, Toma PH (2007) PQRI Recommendations on particle-size analysis of drug substances used in oral dosage forms. J Pharm Sci 96(6):1451–1467

Nichols G, Byard S, Bloxham MJ, Botterill J, Dawson NJ, Dennis A, Diart V, North NC, Sherwood JD (2002) A review of the terms agglomerate and aggregate with a recommendation for nomenclature used in powder and particle characterization. J Pharm Sci 91(10):2103–2109

Waterman KC, Carella AJ, Gumkowski MJ, Lukulay P, Macdonald BC, Roy MC, Shamblin SL (2007) Improved protocol and data analysis for accelerated shelf-life estimation of solid dosage forms. Pharm Res 24(4):780–790

Author Biography

Dr. Jon V. Beaman is a Senior Director, Analytical R&D, Worldwide Pharmaceutical Sciences, and is based in Sandwich, Kent. He has 15 years experience in a number of analytical roles including support for a wide variety of APIs and drug products at all stages of development, setting up and running a pilot plant analytical support unit, running a separation sciences, being responsible for an analytical group supporting all aspects of the veterinary medicine portfolio covering a large number of products and dosage form types and running the Pfizer Global Stability group. Dr. Beaman studied for his PhD in chromatography/mass spectrometry with Professor Dai Games at Swansea.

Chapter 27

Setting Specifications for Drug Products

Abbie Gentry

Abstract Developing specifications for global drug products requires a multi-faceted approach that begins with an understanding of the attributes of the formulation, and how to assess those attributes that are critical to the safety, efficacy and quality of the final drug product. An understanding of global regulatory requirements, as well as requirements that are specific to significant markets adds additional complexity. In addition to developing specifications for the drug product to meet through shelf-life, many drug products require the development of release specifications which require an understanding of stability trends and batch-to-batch variability. This session will include several examples of how these approaches were used.

1. Introduction

To reliably manufacture and release a drug product, it is important to understand the critical attributes of the product and how to control those critical attributes, whether the control is through manufacturing process or through the incoming materials. Satisfactory control of critical product attributes is confirmed by product release and stability testing, as shown by results that meet the acceptance criteria for the drug product. There are three steps followed in setting specifications: the first one is to identify what are the critical product and process attributes in the product, the second step is to evaluate regulatory requirements, and the third step is to evaluate stability trends of the product. Several examples are used to illustrate the concepts presented.

2. Understand Critical Product Attributes

Assuring a quality product requires understanding, control, and confirmation of critical product attributes. An example of an oral solution will be used as a basis for an initial discussion, but the same principles apply to other dosage forms. A high level map of the process from manufacturing, packaging, distribution, and dosing the patient is shown in Fig. 27-1. Each step of the process needs

From: *Pharmaceutical Stability Testing to Support Global Markets*: Pharmasp,
Edited by: K. Huynh-Ba, DOI 10.1007/978-1-4419-0889-6_27,

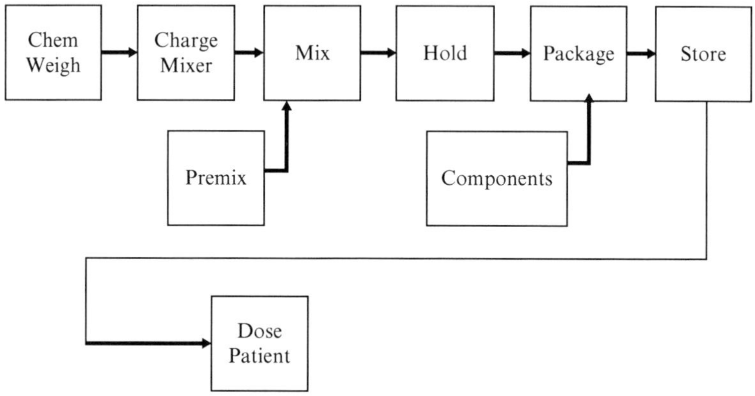

Fig. 27-1. Schematic diagram of a process for an oral liquid dosage form from manufacturing to patient dosing

to be considered when developing specifications and controls for the drug product.

Each step should be evaluated for critical attributes that may have an impact on the safety, efficacy or quality of the drug product.

One of the early steps in the manufacturing process for a liquid oral dosage form is the addition of a thickener or suspending agent. This is often a critical process parameter, as appropriate shear may be necessary for the dispersion and hydration of a suspending agent to assure that the active pharmaceutical ingredient is uniformly dispersed, and does not settle. The primary source of control would be appropriate process controls such as mix speed and mix time. In addition, there are quality control tests such as viscosity or resuspendability that can be used to confirm satisfactory hydration of the suspending agent.

General specifications are also developed to assure that the patient receives accurate dosing of the proper medication. To assure the proper medication, identification, product description – appearance, color, odor, and specific markings (for solid dosage forms) are typical specifications. To assure an accurate dosage to the patient, assay, uniformity, dissolution, and resuspendability may be the important tests.

Purity of the drug product is confirmed through studies such as chromatographic tests for process impurities and degradation products, microbial limits tests, and assays for antimicrobial and antioxidant preservatives.

The use of antimicrobial preservatives needs to be justified through a safety assessment and antimicrobial efficacy testing. Antimicrobial preservatives are not an alternative to appropriate manufacturing controls. Microbial limits are established by considering the nature of the product: for example, if it is a natural product, typically salmonella is evaluated; if it is an oral solution or suspension, *E. coli* is evaluated because it is an oral pathogen; if it is a rectally, urethrally or vaginally administered product, yeast and molds are evaluated. Specific indicator organisms that are used in evaluating antimicrobial effectiveness include: *E. coli*, *P. aeruginosa*, *S. aureus*, *C. albicans*, *A. niger*.

In addition to tests on the particular dosage form, there may be considerations concerning the packaging or a dose delivery device. These considerations include extractables and leachables from the packaging and labeling components, and specific requirements such as uniformity of doses from the device. Product attributes that may affect delivery of an accurate dose from a package or a dose delivery device include resuspendability, viscosity and particle size distribution.

Finally, there are requirements to prevent small children from accessing medication from a package, which is covered by child-resistant requirements in the United States.

3. Understand Regulatory Requirements

The definition of a specification according to FDA Guidance for Industry Changes to an Approved NDA or ANDA (2004) "is a quality standard (i.e., tests, analytical procedures, and acceptance criteria) provided in an approved application to confirm the quality of the drug substances, drug products, intermediates, raw materials, reagents and other components, including container closure systems and in-process materials." This guidance also indicates that acceptance criteria "are numerical limits, ranges, or other criteria for the tests described. Conformance to a specification means that the material when tested according to the analytical procedures listed in the specification will meet the listed acceptance criteria." The critical point is that the acceptance criteria alone are not the specifications, but the specifications are a combination of the analytical methods and the acceptance criteria. If a new analytical method is used to test a product, and that method is not equivalent to the original method, the results from the new method may not be equivalent to those from the original method, and the original acceptance criteria may no longer be important.

The critical product attributes are those which affect safety, efficacy, and quality, thus these define specification requirements. Examples include physical requirements such as control of polymorphs and particle size; chemical requirements such as active pharmaceutical ingredient and stereoselective assays, content uniformity, and impurity tests; microbiological requirements such as total aerobic count, and preservative content. In addition, specific functional tests such as dissolution, disintegration, and preservative efficacy may need to be considered.

Many regulatory authorities have developed a significant body of information relative to specifications for active pharmaceutical ingredients as well as finished products. A partial list of a few of the important guidances include FDA Guidance documents, ICH Guidance documents, Canadian requirements, Australian requirements, EMEA Scientific Guidances, and WHO requirements. Up-to-date versions of many of the harmonized documents can be found on the FDA, EMEA, WHO websites. Country-specific requirements can usually be found on the websites of the specific regulatory authority. A list of many of the international health organizations and regulatory agencies can be found on the FDA website at http://www.fda.gov/oia/agencies.htm.

These documents include general guidance documents that impact setting specifications with topics as wide ranging as impurities, genotoxic impurities, residual solvents, stability, and in-use stability. In addition, there are documents for specific cases such as nasal delivery systems, liposomes, oral disintegrating tablets, oral liquid and solid dosage forms. Many regions also have compendia requirements which include general chapters such as dissolution, heavy metals, and specific monographs for active pharmaceutical ingredients. Some pharmacopeia such as the United States Pharmacopeia also include requirements for specific individual drug products. A list of pharmacopeias can be found on the WHO website: http://www.who.int/medicines/publications/pharmacopeia/WHOPSMQSM2006_2_IndexPharmacopoeiasUpdated.pdf

As an example of how to apply all of these guidance documents, consider an aqueous solution intended for injection. A list of the parameters to evaluate may include:

- Description
- Identification
- Assay for active pharmaceutical ingredients
- Degradation products
- Preservative content (antimicrobial and antioxidant)
- Uniformity of dosage units (for single use containers)
- Delivered dose uniformity (package dependant)
- Osmolality
- Particulate matter
- Sterility
- Bacterial endotoxins

If the aqueous solution is intended for oral use in a multi-dose package, the specifications would be more limited:

- Description
- Identification
- Assay for active pharmaceutical ingredients
- Degradation products
- Preservative content (antimicrobial and antioxidant)
- Viscosity

The interrelationships between specifications is shown by an example of an evaluation of pH for oral suspensions. Consider a product with a USP specification for pH of 4.4–4.6: This particular oral suspension is typically formulated with a preservative with a p*K*a of about 4.2, and the efficacy of the preservative is typically much better below its p*K*a. A study of preservative efficacy testing demonstrated a 1.8 log reduction in yeast at pH 4.6, and a 3.7 log reduction at pH 4.0. The improvement in preservative efficacy at pH 4 is a good reason to evaluate the buffer system used in the formulation, and determine if the other components of the suspension are stable at pH 4.0.

4. Understand Process Capability

Another example of the interrelationship between specifications and critical product attributes involves an evaluation of pH and process capability. Assuming that all production batches come from the same population, then measurement of these batches are described by the normal distribution as shown in Fig. 27-2. The characteristics of a normal distribution are such that 99.7% of the measurements fall within ±3 standard deviations from the mean.

Process capability describes the relationship between the specifications and the distribution of measurements, as shown in (27-1):

$$\text{Cpk} = \min\left[\frac{\text{USL}-\mu}{3\sigma}, \frac{\mu-\text{LSL}}{3\sigma}\right] \quad (27.1)$$

Many manufacturing groups target processes that have a process capability better than 1.33. Figure 27-3 shows a process with a mean of 100, and a standard

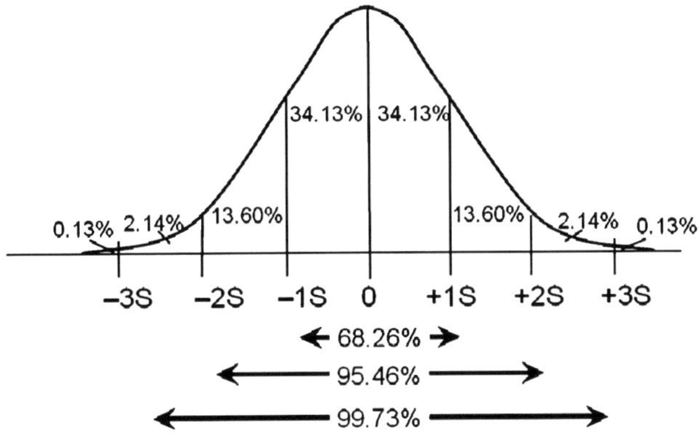

Fig. 27-2. A normal distribution

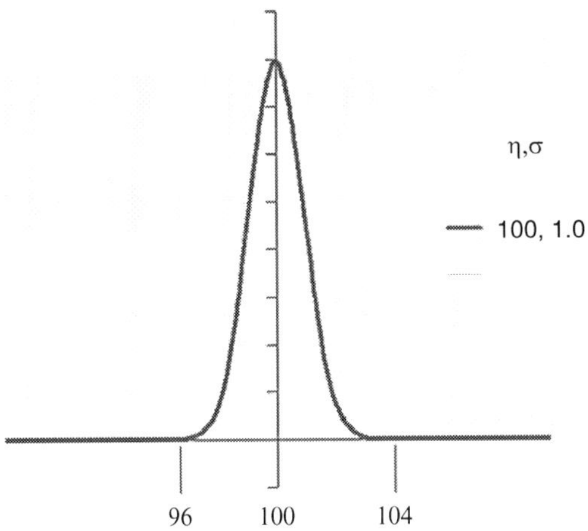

Fig. 27-3. Schematic of a process with Cpk = 1.33

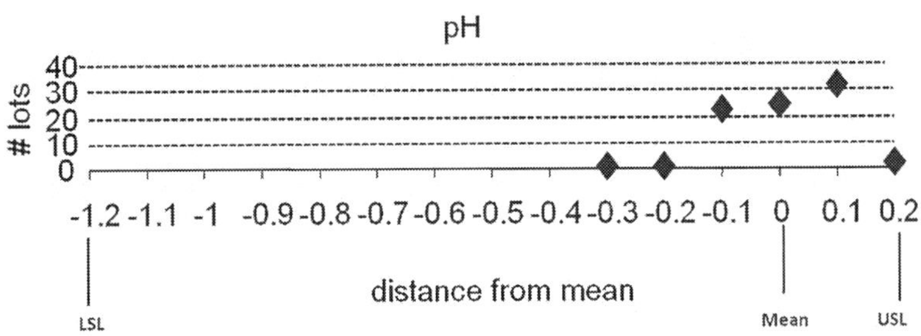

Fig. 27-4. Plot of pH results

deviation of 2. If the acceptance criteria for this process are 96–104, the capability of the process is 1.33.

The process capability case study is a different oral suspension than the one discussed earlier. pH test results of 100 batches are shown in Fig. 27-4. It is obvious that mean pH is not centered within the specification range, and this has a negative impact on the process capability. In this case, Cpk is determined by the distance between the mean of the measurements and the upper specification:

$$\text{Cpk} = \min\left[\frac{\text{USL} - \mu}{3\sigma}, \frac{\mu - \text{LSL}}{3\sigma}\right] = 0.69 \tag{27.2}$$

To improve the process capability, actions should be taken to move the mean of the process to the center of the specification range. Either the specification range can be changed, or the buffer system of the formulation can be changed to move the pH to the center of the specification. In this case, neither approach is straightforward. There are specific aspects of the formulation that impact safety and efficacy that prevent widening the upper end of the specification range. In addition, there are formulation considerations that prevent the use of a buffer system that provides a lower pH. Thus, the best approach to improve process capability is to evaluate the pH test, and reduce the observed variation to less than 0.1 pH unit.

5. Understand Stability Trends

In general, the slope of any stability changes and the expiry period is determined when the 95% confidence limit intersects the specification as shown in Fig. 27-5.

In the first case study, the data from several stability studies was sufficiently different between the lots that each lot was evaluated individually. As shown in Fig. 27-6, the minimum observed result is 94.8, and the 95% confidence limit

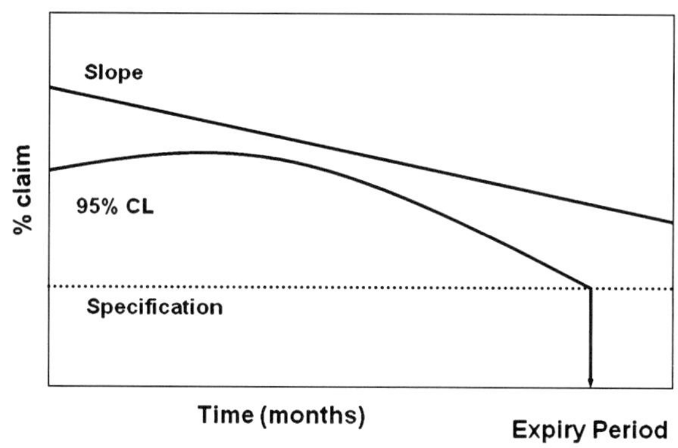

Fig. 27-5. Schematic diagram for evaluating stability data

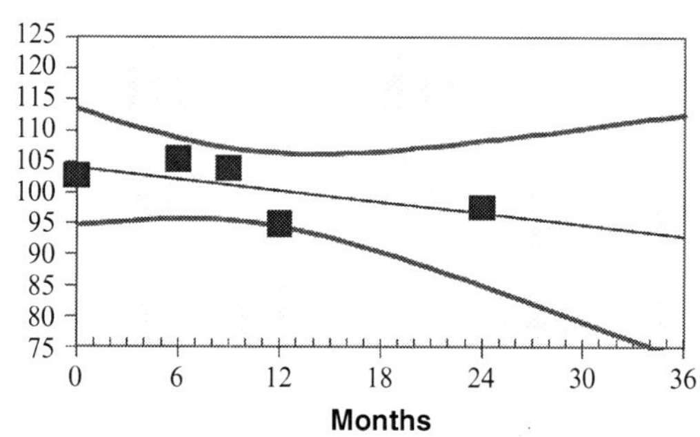

Fig. 27-6. Stability trends for a single lot with significant variability

does not support a 24 month expiry period for this product when specifications of 90.0–110.0 are used. In this case, it is necessary to do one of the following: justify a wider specification range, define a shorter expiry period, or take actions to minimize the spread of the confidence limits by minimizing the variation in the data observed, or adding additional data points.

The next case study summarizes some considerations to setting release specifications. Figure 27-7 shows the stability data for a number of lots where the batches have a common slope when evaluated at a p value of 0.25, but the intercepts cannot be combined. It is clear from this figure that there is no significant slope to the data, and the 95% confidence limit for the common slope is shown by the *solid line* below all the *dotted lines* from each individual lot. The variation in the intercepts was used to establish release specifications for this specific product.

The final example concerns evaluating degradation products are observed in products with multiple active pharmaceutical ingredients. In this case, the drug product contains four active pharmaceutical ingredients and the levels of the active ingredients are very different as are the response factors of the degradation products. Figure 27-8 shows a chromatogram where one of the active pharmaceutical ingredients is less than 0.5% of the most concentrated API.

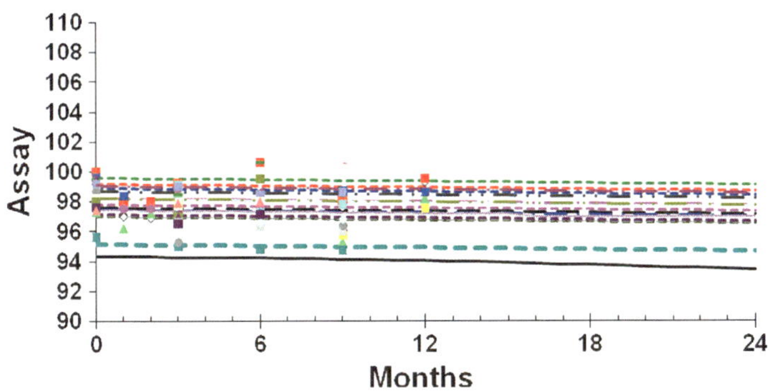

Fig. 27-7. Stability trends for a product with common slope and separate intercepts

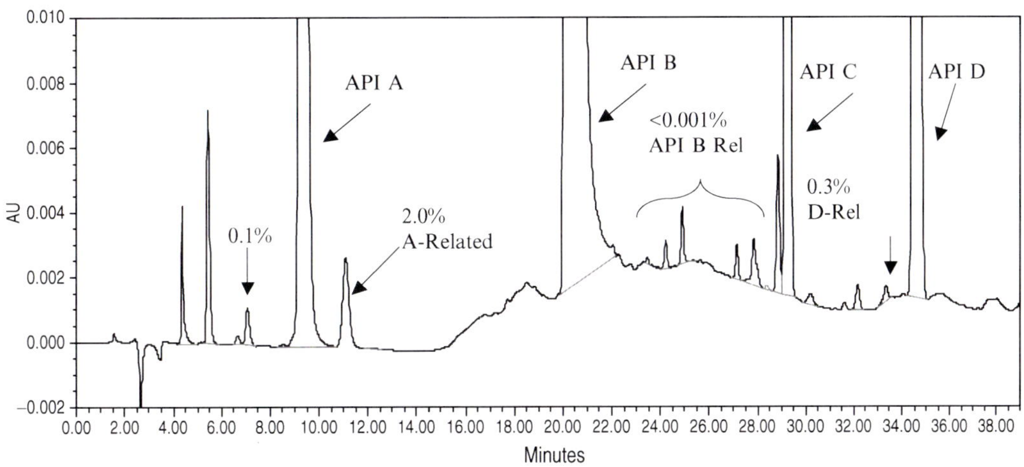

Fig. 27-8. Example chromatogram from drug product with four APIs

For this specific product, it is important to have a well-defined decision-making process for assessing the peaks in the chromatography. Several well resolved peaks represent much less than 0.001% of the API that they are related to. Other peaks with much less area represent 0.3% of a degradation product of another API in the formulation. In this product, unidentified chromatographic impurities are assumed to be degradation products of the lowest concentration API in the formulation, and are reported and identified according to the ICH thresholds for that API. Although some of the other APIs in the formulation may have reporting thresholds that are lower in terms of percent of active, the low concentration of API D assures that the unidentified impurities will be assessed at the lowest level in terms of overall weight in the drug product. As more information is gathered, and it is possible to assign peaks to other APIs through spectral analysis, forced degradation studies, etc, then the relative response factor of the related API is used to evaluate reporting, identification and qualification threshold levels.

6. Summary

In summary, a scientific approach to understanding the critical attributes of a product, and the implications to efficacy and safety lead to a rational approach to setting specifications. The specifications that are developed this way should be evaluated against all applicable regulatory guidance to understand if additional specifications should be considered, or a justification developed for not testing a specific product. Finally, an evaluation of process capability and stability trends leads to the understanding of developing requirements so that the product maintains its quality through shelf-life.

Author Biography

Dr. Abbie Gentry is the Sr. Director of Analytical Technology at McNeil Consumer Healthcare. In this role, she is responsible for oversight of world-wide R&D analytical laboratory practices for over-the-counter drug products. She has over 20 years of experience in directing both analytical laboratories and formulation development groups, with particular interest in stability evaluation and formulation approaches to resolve stability issues. Dr. Gentry has previous experience at Johnson & Johnson Merck, Rohm and Haas, and Rhone-Poulenc Rorer in the areas of analytical method development, stability evaluation, and characterization of polymeric systems. She received her M.S. and Ph.D. in Physical Chemistry from the University of Michigan, and her B.S. in Chemistry from Colorado College.

Chapter 28

Highlights of Investigating Out-of-Specifications Test Results

Saji Thomas and Kim Huynh-Ba

Abstract Investigating out-of-specification (OOS) results and establishing corrective and preventive action (CAPA) are the critical aspects of the operation of a compliant quality control unit. Investigations are routinely reviewed and scrutinized by the regulatory agencies during cGMP inspection. On September 30, 1998, the FDA announced the availability of the draft guidance for "Investigating Out-of- Specification (OOS) Test Results for Pharmaceutical Production." This draft has been the subject of some controversies and much discussion. On October 12, 2006, the FDA published the final guidance. This paper summarizes important changes listed in the new version of the guidance and also the implementation of the guidance in the new Quality by Design (QbD) risk-based approach. A review of potential root causes and the effective approach for establishing CAPA are also discussed.

1. Introduction

21 CFR 211.192 states that the failure of a batch or any of its components to meet any of its specifications shall be thoroughly investigated, whether or not the batch has already been distributed. The investigation shall extend to other batches of the same drug product and other drug products that may have been associated with the specific failure or discrepancy. A written record of the investigation shall be made with conclusion and follow-up.

This paper merely summarizes the Out-of-Specification (OOS) investigation process based on the FDA guidance and gives some thoughts on how the root causes can be identified. For more information, one should consult the FDA Guidance directly.[1] Subbarao and Huynh-Ba also discussed the mechanism of an OOS investigation as well as an evaluation of stability data (Huynh-Ba 2008).

[1] FDA Guidance to the Industry, "Investigating OOS test results for pharmaceutical production", (1996).

From: *Pharmaceutical Stability Testing to Support Global Markets*: Pharmasp,
Edited by: K. Huynh-Ba, DOI 10.1007/978-1-4419-0889-6_28,

2. Historical Considerations

2.1. The Barr Decision

The most important aspect of Barr decision was that the methods and the process were not validated. The key testing, resulted that the decision was obtained and the out of specification (OOS) result was discarded.

The ruling settled much of the debate between the industry and the FDA concerning lab investigations, raw data handling, method validation and data integrity. PhRMA has published an article discussing this topic as part of their Acceptable Analytical Practices series.[2]

2.2. Judge Wolin's Legal Ruling-Issues

The Barr case has concluded several non-compliance practices:

- Averaging OOS values with "in specification" values to get passing results.
- Ineffective program for process validation.
- Discarding raw data.
- Conducting multiple retesting with no-defined endpoint.
- Performing inadequate failure investigation.
- Lack of method validation.

His decisions indicate that the history of the product should be considered, the exact cause is difficult to determine and the use of outliers is restricted. All companies should have a retest policy and a protocol that define the point in which testing stops and the data evaluation occurs.

3. Scope of the OOS Guidance

This guidance applies to chemistry-based laboratory testing of drugs regulated by CDER. It is directed toward traditional drug testing and release methods. These laboratory tests are performed on active pharmaceutical ingredients, excipients and other components, in-process materials, and finished drug products.

3.1. Assessing OOS Test Results

The source of OOS results should be identified as either an aberration of measurement process or as manufacturing process. If a batch is rejected based on an OOS result, the investigation is to determine if the result is associated with other batches of the same drug product or other products. In addition, batch rejection does not negate the need to perform the investigation and data should be reviewed thoroughly, timely, unbiased, well-documented, and scientifically sound.

The contract laboratory should convey its data, findings, and supporting documentation to the manufacturing firm's quality control unit (QCU), who should then initiate the full-scale OOS investigation.

3.1.1. System Performance Check Standard

If reference standard responses indicate that the system is not functioning properly, all of the data collected during the suspect time period should be

[2] PhRMA Acceptable Analytical Practices: Investigating OOS results (2004).

properly identified and should not be used. The cause of the malfunction should be identified and, if possible, corrected before a decision is made whether to use any data prior to the suspect period.

3.1.2. Obvious Errors

If errors are obvious, such as the spilling of a sample solution or the incomplete transfer of a sample composite, the analyst should immediately document what happened, and thus data can be invalidated and the test can be repeated due to analytical errors. Analysts must not knowingly continue an analysis that they expect to invalidate at a later time for an assignable cause.

3.2. Supervisor's Responsibility

The supervisor's assessment should be objective and timely. There should be no preconceived assumptions as to the cause of the OOS result.

Data should be assessed promptly to ascertain if the results might be attributed to laboratory error, or whether the results could indicate problems in the manufacturing process.

3.3. Responsibility of Laboratory Management

Laboratory management should ascertain not only the reliability of the individual value obtained, but also the significance these OOS results represent to the laboratory quality assurance program. The firm's upper management should appropriately monitor these trends and ensure that any problematic areas are addressed.

3.4. Root Cause Analysis

The root cause of an OOS result must be determine. The OOS result has to come from any of five criteria; it could be due to the *methods* (e.g., changes to the method, chemist observations, validation report), it could relate to the *materials* (e.g., solvents, vials, changes of gasses), *environment* (Temperature and humidity), *equipment* (e.g., detector, injector, pH meter) or *people* (see Fig. 28-1).

3.5. Laboratory Error and CAPA

Laboratory error should be relatively rare. Frequent error suggests lack of laboratory controls. Problem might be due to inadequate training of analysts, poorly maintained or improperly calibrated equipment, or just careless work. Therefore, whenever laboratory error is identified, the firm should determine the source of that error and take corrective action to prevent recurrence. To ensure full compliance with the cGMP regulations, the manufacturer also should maintain adequate documentation of the corrective action.

4. Phases of Investigations

4.1. Phase 1 of the Investigation

When the initial assessment does not determine that laboratory error caused the OOS result and testing results appear to be accurate, the main objective is to identify the root cause of the OOS result and take appropriate corrective and preventative action.

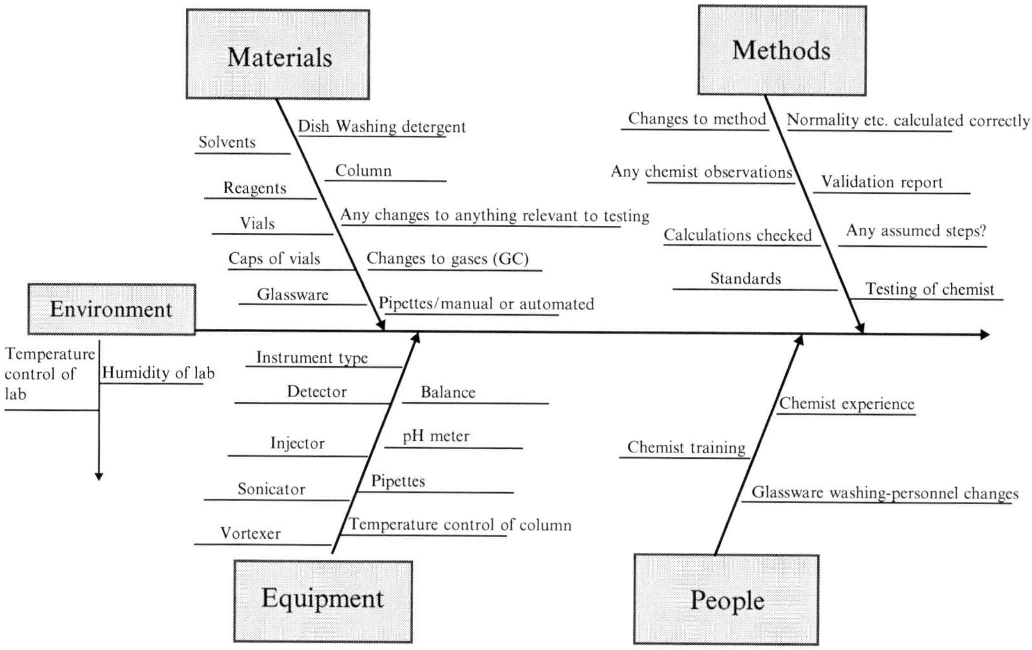

Fig. 28-1. Root cause analysis fishbone diagram

This phase should go to a full scale investigation that would include a review of production and sampling procedures, additional laboratory testing and the impact of OOS result(s) on already distributed batches.

4.2. Phase 2 of the Investigation

This phase of the investigation would be conducted by the QCU. Other departments such as manufacturing, process development, maintenance and engineering, must be involved in the investigation at this point.

5. Review of Production

A full-scale OOS investigation should consist of a timely, thorough, and well-documented review that would include the following:

- A clear statement of the reason for the investigation.
- A summary of the aspects of the manufacturing process that may have caused the problem.
- The results of a documentation review, with the assignment of actual or probable cause.
- The results of a review made to determine if the problem has occurred previously.
- A description of corrective actions taken.

If the OOS investigation confirms the OOS result and is successful in identifying its root cause, the OOS investigation may be terminated and the batch can be rejected. The failure investigation will then extend to other batches or products that may have been associated with this specific failure.

6. OOS Follow Up

OOS result may indicate a flaw in product or process design. For example, a lack of robustness in product formulation, inadequate raw material characterization or control, substantial variation introduced by one or more unit operations of the manufacturing process, or a combination of these factors can be the cause of inconsistent product quality. In such cases, it is essential that the redesigning of the product or process has to be undertaken so as to ensure the reproducible product quality.

6.1. Retesting

The sample used for the retesting should be taken from the same homogeneous material that was originally collected from the lot, tested, and yielded the OOS results. It is often important for the predefined retesting plan to include retests performed by an analyst other than the one who performed the original test.

A second analyst performing a retest should be at least as experienced and qualified in the method as the original analyst.

FDA inspections have revealed that some firms use a strategy of repeated testing until a passing result is obtained, then disregarding the OOS results without scientific justification. This practice of "testing into compliance" is unscientific and objectionable under cGMPs.

The maximum number of retests to be performed on a sample should be specified in advance in a written standard operating procedure (SOP). The number may vary depending upon the variability of the particular test method employed, but should be based on the scientifically sound principles. The number of retests can not be adjusted depending on the results obtained. The firm's predetermined retesting procedures should contain a point at which the additional testing ends and the disposition of the batch is evaluated.

6.2. Reporting of Results

In the case of a clearly identified laboratory error, the test will be repeated. Retest results would substitute for the original test result.

If no laboratory errors are identified in the first test, there is no scientific basis for invalidating initial OOS results in favor of passing retest results. All test results, both passing and suspect, should be reported and considered in batch release decisions.

6.3. Case Studies

6.3.1. First Case Study

For the first case study, the initial result was 89% while the specifications restrict the results to 90–110%. The first retest result was 90% and the second was 91%. Even though the average of these two retest results is within the restricted specifications, this batch cannot be released, because the retest results actually confirms the initial results as it is within the accuracy and precision of the instruments used in the tests.

6.3.2. Second Case Study

For the second case study, the initial result was 80% while the specification is 90–110%. The retest results gave 85% and 105%. It's obvious that these results are too far apart which would indicate variability in the sample size or a problem with the instruments used. In both cases, this batch cannot also be released immediately without farther investigation for the cause of the variability in the results.

6.3.3. Result Reporting

Each sample should generate one result. For several samples, results must be averaged to one single result representing them all. The use of replicates to arrive at a single reportable result, and the specific number of replicates used, should be specified in the written, approved test method. Acceptance limits for variability among the replicates should also be specified in the method, while unexpected variation in replicate determinations should trigger remedial action as required by § 211.160(b) (4).

The quality control unit (QCU) is responsible for interpreting the results of the investigation. In those instances where an investigation has revealed a cause and the suspect result is invalidated, the result should not be used to evaluate the quality of the batch or lot.

A confirmed OOS result indicates that the batch does not meet established standards or specifications and should result in the batch's rejection, in accordance with § 211.165(f), and proper disposition. In cases where an investigation (1) does not reveal a cause for the OOS test result and (2) does not confirm the OOS result. The OOS result should be given full consideration in the batch or lot disposition decision.

7. Outlier Test

Outlier testing is a statistical procedure for identifying from an array those data that are extremes. It may result from deviation from prescribed test method or variability in the sample. The possible use of outlier test should be determined in advance in an SOP or an investigation protocol. For biological assays where a high variability may exist, the use of an outlier test may be appropriate. However, for validated chemical assays with small variance, the use of outlier test is not encouraged. Outlier tests are not applicable for testing such as content uniformity or dissolution, where the variability of the product is being assessed.

8. Typical Number of Replicate Retests

Mathematically speaking, the standard deviation (SD) = $\sigma/\sqrt{n}$, where σ is the standard deviation, n is the number of replicate tests and SD can be used to represent the precision of measurements as the precision is inversely proportional to the standard deviation. Therefore, the precision of the measurements increases with the number of retests being conducted. Typically, the number of tests was chosen to be 8, which was the case in the Barr decision. The number of retest may vary due to the variability of the particular test method; however, it should be based on scientifically sound principles.

9. Averaging the Results

In cases where a series of assay results (to produce a single reportable result) are required by the test procedure and some of the individual results are OOS, some are within specification, and all are within the known variability of the method, the passing results are no more likely to represent the true value for the sample than the OOS results. For this reason, a firm should err on the side of caution and treat the reportable average of these values as an OOS result, even if that average is within specification.

Filed alert report (FAR) should be submitted within 3 working days. Unless the OOS result on the distributed batch is found to be invalid within 3 days, an initial FAR should be submitted. A follow-up FAR should be submitted when the OOS investigation is completed.

10. Preventing OOS Results

10.1. Quality by Design

The first and main step in preventing OOS results is to understand the process from a manufacturing perspective. That would include a good understanding of the physical and chemical attributes of the Active Pharmaceutical Ingredient (API), and an understanding of the critical process parameters and the critical formulation attributes.

It's also important to do a statistical design process validation, this way it would be easier to actually pinpoint the process parameters and put additional controls to identify the cross parameters and also the quality attributes that are important to the manufacturing of the drug product.

10.2. Analytical Method

The methods used should be validated based on the current ICH guidance. Critical parameters of the method have been established by statistically designed robustness studies.

Method uses should also be Quality Control (QC) friendly. For an example, there should be a small amount of accepted tolerance to ease the controlling of the overall quality. Transferring the method to the QC, using co-validation as a technique, should give the QC a better understanding of these methods at early stages.

Another important aspect is the real time quality control. Having control of the system step-by-step should make it easier to pinpoint any problems and

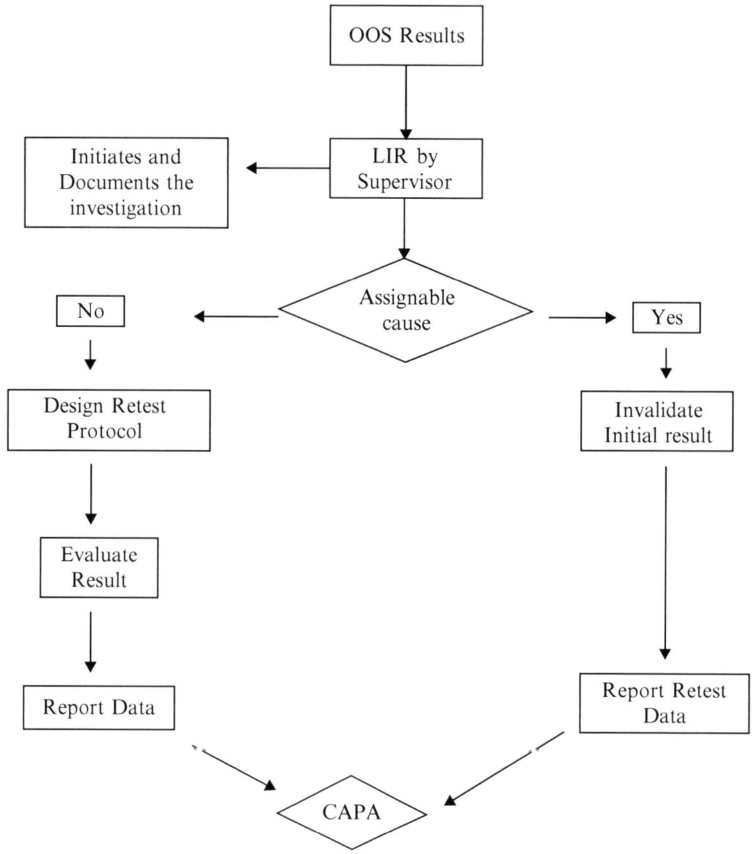

Fig. 28-2. OOS flowchart

work on solving it, rather than inspecting the end product for problems and then back trace it to find its cause.

10.3. Quality Systems

There should be a Standard Operating Procedure (SOP), because a well established procedure is a key to prevent OOS. It is recommended that each firm should develop three sets of SOP's:

- Out-of-Trend (OOT).
- OOS for developmental products.
- OOS for marketed products.

10.4. Stability Study

A stability study should cover appropriate storage conditions and packing configurations, evaluate all data for all attributes, and testing completed in a reasonable timeframe. OOS result of stability testing must go through a similar investigation process as for release testing.

11. Conclusions

As soon as the obtained of specifications results are obtained, the supervisor fills out a laboratory investigation report and then look for assignable cause. If there is no assignable cause, retesting should take place to evaluate the result, report the data and establish a corrective and preventive action. If there was an assignable cause, it's necessary to invalidate the initial results and report the retest result. (See Fig. 28-2).

Reference

Huynh-Ba K (ed) (2008) Handbook of stability testing in pharmaceutical development: regulations, methodologies, and best practices. Springer Science + Business Media, New York, pp 269–275

Author Biographies

Mr. Saji Thomas joined Par Pharmaceuticals as Associate Director of Quality Control Operations. His responsibilities at Par include raw material and finished product release, process validation, lab automation and technical services. Prior to joining Par, he worked at Barr laboratories for 2 years as Manager of Analytical Services. Before that, he was at Purdue Pharma for 12 years, where he held positions of increasing responsibility, including Group Leader of Chemical Development and Group Leader of Analytical R&D. Saji has organized and moderated several workshops and symposia for AAPS. He has been an invited speaker on a variety of CMC topics for pharmaceutical workshops and conferences in both the US and Europe. Saji holds B.Sc. and M.Sc. degrees in chemistry from the University of Kerala.

Mrs. Kim Huynh-Ba is the Technical Director of Pharmalytik. She has over 24 years of experience in various analytical areas of pharmaceutical development and a primary focus in stability sciences, analytical development, outsourcing and technology transfer management. She has authored numerous technical publications and is invited frequently to present at national and international conferences. She is the editor of the "Handbook of Stability Testing in Pharmaceutical Development: Regulations, Methodologies and Best Practices," Springer.

Chapter 29

Strategies for Ensuring Regulatory and cGMP Compliance of Outsourced Stability Programs

Eda Ross Montgomery

Abstract The process of outsourcing requires continuous improvement and open communication. In this article, the author examines the stages of outsourcing, beginning with the decision to outsource and select the vendor. Following the topic of choosing the vendor, the article describes the startup phase and offers advice on how to monitor the ongoing studies and manage change during an ongoing stability study or a development program. During the management-of-change phase, the author illustrates relevant, real-life examples from her career of what to avoid during the management of stability studies. The article concludes with the topic of continuous improvement.

1. Introduction

Devising the right strategy to ensure regulatory and cGMP compliance stability programs requires thorough planning and ongoing management of time and resources.

2. The Decision to Outsource

Most companies will quickly make a decision to outsource based either on efficiency, high-volume testing on a specific formulation or a specific product, or a short-term or medium-term need for an increasing capacity that they just do not have internally. However, it may make sense to outsource if there is a need for specialized equipment. For example, if a company is working on a project that utilizes LCMS as the detection mode but is not prepared to invest a large amount of financial resources in LCMS equipment for a given product early in development, it would be reasonable to outsource. Another situation where outsourcing may be needed is when there are predictable changes to methods, specifications, or the formulation. However, with proper management, a company could outsource at any point in development.

From: *Pharmaceutical Stability Testing to Support Global Markets*: Pharmasp, Edited by: K. Huynh-Ba, DOI 10.1007/978-1-4419-0889-6_29,

3. Choice of Vendor

A critical part of the outsourcing process is selecting the vendor. Although the most common determinant is cost, this should not be used as the primary factor because one will miss the opportunity to pick the vendor who provides the most value. Stability programs, when they are not carefully managed, can quickly escalate in cost. The phase of development is an important factor when selecting a vendor, since a given product at phase I could be outsourced to the same specialty vendor for the entire development lifecycle, or manufacturing, release and stability campaign(s) could be done at the same vendor. The number of projects and the number of vendors to manage are also factors to be taken into consideration, as there is a tradeoff between flexibility due to changes in capacity vs. the challenge of integrating multiple vendors and quality/business processes. Equally important is the fit of the quality system of the vendor with that of the company which is related to the degree of oversight needed. Last is company culture, which is most often observed by geographic location. For example, if a company is working with a vendor 12 time zones away, a product with a low rate of change should be selected for the first project. If extensive troubleshooting is needed, it will be much more difficult to quickly resolve problems.

Technical assessments are the vital parts in choosing a vendor. An on-site evaluation, either in conjunction with or in partnership with the Quality organization is highly recommended. During the evaluation, significance is given to consider the number, capacity, and diversity of the vendor's stability chambers, the types of their chromatographic equipment, detectors, and size of the lab. The size of the organization – not only the number of analysts available to work on a given project but the number of quality assurance personnel who will review the company's data is also an important factor. Depending on the complexity of the project(s), it is a good idea to assess the education and experience of the vendor's quality assurance (QA) personnel in order to understand the extent to which they are going to be able to assist in troubleshooting, method development, issue resolution and development of strategies to comply with ICH or FDA guidelines. One area where they could add significant value is in helping the sponsor company to develop a strategy to deal with a potentially problematic issue that could come up during a pre-approval inspection.

Once a technical assessment has been done, one can delve into the less concrete matters, such as evaluating what the right vendor size is relative to the company size and vice versa. There are pluses and minuses to either approach. The advantage to favoring a large vendor relative to the sponsor company is that the larger vendor has well-established practices. In general, their systems are experienced with audits, and their staff and facilities have adequate capacity. However, these practices and systems may be more well-established, posing the potential for them to be more rigid and offer less flexibility on timelines and/or testing. A smaller vendor might have the ability to be very responsive to sponsor requests because they tend to have a high degree of ownership of the company's project; however, changes in smaller vendors (buyouts, personnel) have a much greater impact on their interactions with the customer. Another aspect to consider is the number of projects vs. the number of vendors to be managed. For projects with diverse needs (i.e., tablet, oral solution, and sterile product) equipment and/or skill set constraints may exist at a single vendor. There is also the potential for limited capacity for some specialized testing done during late development, such as microbial limits testing, or insufficient storage capacity for large stability studies with several packaging configurations. While choosing one vendor per stability study would eliminate many of these risks, it creates other problems. Vendor management of multiple transfers, multiple quality systems, multiple systems for information exchange and multiple change controls results in some degree of duplication. This can lead to a greater risk of, for example, two sets of inconsistent results on the same samples.

Common advice associated with outsourcing stability testing is "*Pick any two: Speed, cost, and quality*." If speed is the priority, electronic systems for data and information exchange are usually preferred. One might prefer the cost benefits that come from centralized services with one vendor. From a quality standpoint, the company should determine up-front the frequency and type of interactions. Do they prefer a vendor with high customer contact and regular status updates on projects and investigations or a contract-and-deliver approach where one does not hear from the vendor until a finalized report has been issued? The last point to consider in respect to vendor selection is deciding between a manufacturing site with development capabilities and a development site with manufacturing capabilities. The former is the best suited for routine manufacture and is likely to have robust systems, especially in project management, and the latter is more flexible in timelines and prepared for changes in project scope.

4. The Startup Phase

After the vendor has been selected, the project kickoff phase begins. The first part of this phase is building the systems to be used. Even with the best vendors, it's important to build a solid base and ensure that one has confidence in the data and the way it is generated, documented and reported. One should also be aware of the three different types of vendor relationships – fee for services, FTE agreements, and long-term on-site relationships, and select the option that provides the degree of oversight that matches the risk the company is willing to take in potential program delays. If the intent is to minimize the oversight, it is highly recommended that the data used to establish or revise storage conditions or retest date, and OOS results have to be reviewed. It is also essential to agree on mutual expectations for activities as simple as agreeing on who will approve vendor documents from both the vendor and the sponsor company, and to establish a process to ensure whether the documents are integrated into both companies' change control systems. Vendor lead times for introduction of new methods can also vary greatly, depending on whether the sponsor's method is used as provided; the sponsor's method is integrated into the vendor's system, or something in the middle. Having the understanding of the mechanics of the process and its duration can prevent unforeseen delays. Furthermore, when establishing expectations for communication of unexpected results, one should agree on a common definition of unexpected results along with specifics on how investigations are conducted. If the sponsor expects the vendor to notify them at each step in the investigation, both companies should have this understanding at the project outset to avoid delays in closing investigations. Finally, the means by which stability data is transferred – paper

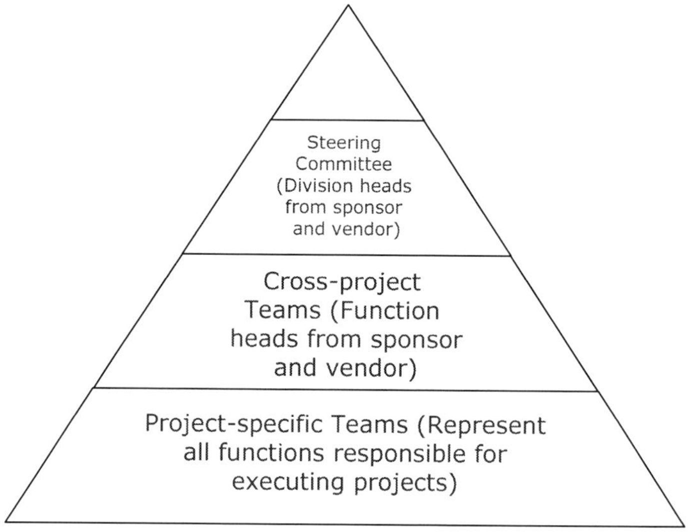

Fig. 29-1. Process for governing stability studies and managing inter-company relationships

or electronically – and whether these data are editable or locked can have an enormous impact on the data review, trending, and preparation of regulatory submissions at the sponsor if the systems are not matched in a way that allows efficient use of resources at both sponsor and vendor. In addition to establishing expectations, the means for ongoing evaluation of the vendor should be communicated up-front. Likewise, it is a good idea to request feedback from the vendor on how they are evaluating the project.

In addition to project-specific startup activities, having a specific process for managing activities with the vendor is important. This process can be pictured as a triangle (see Fig 29-1). At the bottom of the triangle is the project team, which is responsible for day-to-day operations and executing the project deliverables. The next level up is the cross-project team, generally composed of management with departmental or cross-project responsibility. Their role in this process is to discuss cross-project activities that are not working well, systems integration issues, and relationship issues. At the top of the triangle is the steering committee, which gets involved when issues cannot be resolved at lower levels. Even if there are no regular meetings with this group, it is very helpful to establish this group in advance. From inside an organization, it might be obvious at first sight whom to turn to for a particular deliverable, and completely non-obvious to somebody looking in from the outside.

5. Monitoring Ongoing Studies

There are a number of mistakes to be avoided when conducting and monitoring ongoing studies. Preventing mishaps from happening by careful planning followed by prudent monitoring of studies can save a sponsor company substantial time and resources. Table 29-1 lists down some examples of common problems and mechanisms for preventing these problems from occurring.

Table 29-1. Top ten problems when monitoring ongoing studies.

Problem	Mechanism to avoid
Lack of established mechanism for reporting results	Provide data table suitability for regulatory filings
	Predefine results format and determine whether electronic or hard copy will be provided
	Describe results reported in the method or in project transfer documentation (average of duplicate sample preparations, define reporting limit for impurities, process for labeling unknown impurity peaks)
Results not widely available at sponsor lab	Ensure publication process exists for the data
	Provide summary and interpretation of results
	Identify effected documentation (i.e., retest dating, regulatory submissions) and update as needed
Incompatible data transfer/ reporting systems	Identify common reporting tools (software, documentation system, inventory control system), agree on mechanism for transfer and define roles and responsibilities for the sponsor and the vendor for this process
Method not documented completely	Define injection sequence in the method, specifying the frequency of blank injections and standards
	Update method frequently to add detailed instructions or update critical factors
	Document critical factors (i.e., instructions for shaking, specific instrument settings or integration instructions) where they are known and provide guidance on how to avoid them
Investigations not timely (sample solutions discarded before results are approved, OOS discovered after all testing complete, result not considered OOT)	Provide specific instructions on procedure to be followed for OOS and OOT investigations, measure performance and report metrics on timing for each step of investigation at project-specific and cross-project teams
	Regularly and promptly review data and highlight OOT results
Poor turnaround time	Solicit and implement feedback on ways to improve method and project efficiency, ensure all effected personnel are trained on sponsor requirements
Lack of adequate resources in lab or QA	Monitor turnaround time in sufficient detail to highlight resource constraints and inefficiencies (include data review and documentation)
Inadequate sponsor stability program management	Review all studies periodically and cancel redundant or obsolete studies
Changes not communicated in time to efficiently implement them	Ensure the process and turnaround time for implementation of methods are built into project plans and potential updates communicated regularly
Lack of clear communication of expectations	Define sponsor vs. vendor role in investigations and sponsor review and approval process
	Solicit feedback and implement improvements to processes for making changes to project scope, priority, etc.
	Establish, communicate, and regularly report performance metrics at both vendor and sponsor

6. Management of Change

In addition to monitoring the ongoing studies, managing change can pose difficulties in outsourcing. There are two main types of change that occur frequently during stability programs – company growth, either by the vendor or the sponsor, and integrating and managing new products. It's important to have

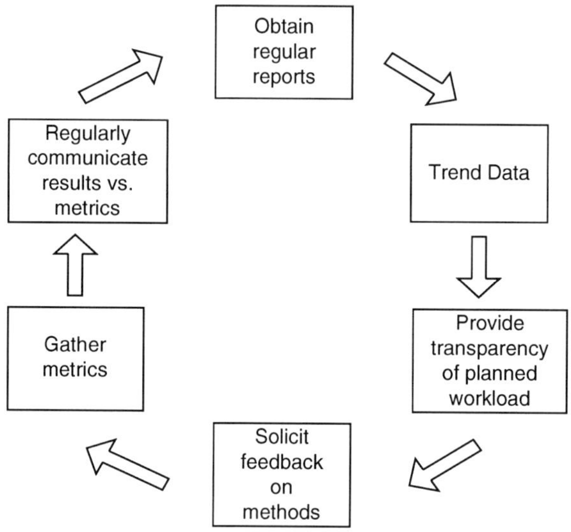

Fig. 29-2. Continuous improvement – maximizing the success of your outsourced stability program

transparency when changes in roles and responsibilities occur and to continue to get feedback on what is and is not working. Areas where the impacts of company growth are most often felt are in the knowledge management tools (small-scale, server or document-based transfer of data to LIMS systems that import data, trend, prepare data summaries suitable for regulatory submissions, etc.), as well as in a company's organizational structure. It's common that, as a company begins to grow, a role that was solely managed by one person is now divided among several people. In this case, the sponsor company needs to recognize the effect of this change on their vendor relationships and ensure that the communication flows are maintained. In this case, designating one company "voice" to direct investigations or coordinate data evaluation can be beneficial. Likewise, whenever changes happen with contract lab growth, the roles and responsibilities of those involved should be reconfirmed, and the composition of each of the cross-project teams depicted in Fig 29-1 should be revisited to ensure the correct composition of each team. For changes in project scope, due to the introduction of new products (i.e., new dosage forms and line extensions), a subset of the vendor selection process should be done.

7. Continuous Improvement

The key to a successful stability program is pictured in the continuous improvement cycle depicted in Fig 29-2. Regular reports, trending, and ongoing communication about method performance and performance metrics are used to identify changes needed to the processes and provide a forum for continuously improving processes, methods, and systems. If done effectively, continuous improvements in the relationships are also realized which further feed the improvement cycle.

Author Biography

Dr. Eda Ross Montgomery received her Bachelor of Science in Chemistry from Rensselaer Polytechnic Institute and her Ph.D. in Analytical Chemistry from the University of North Carolina at Chapel Hill under the direction of Dr. Royce Murray. Eda has over 20 years of experience in the pharmaceutical industry, where she has led CMC teams for all Chemistry, Manufacturing, and Control activities and Analytical Development activities for four NDA filings. She has also headed departments of many aspects of drug development, including Analytical Development, Lifecycle Management, and Commercial Support. She is currently responsible for the implementation of Quality by Design at Vertex, including development of a strategy for the CMC filing, implementation at manufacturing sites, launch, and continuous improvement of Vertex' commercial products.

Chapter 30

Building and Developing of Relationships with Third Party Laboratories

Rob Aitchison

Abstract The key to successful outsourcing to third party laboratories is the relationship, trust and communication between both parties. How you go about building and maintaining that relationship can be dependant on a variety of factors? Why have you chosen to outsource? What outsourcing model are you going to employ? What experience do you have of the specific vendor? However, there are factors that should be considered irrespective of these variables. This paper explores the key aspects of outsourcing stability testing to contract laboratories. This will include looking at the reasons why one might choose the outsourcing option, the potential contracting models to choose, identifying and managing risk and how to develop and maintain good working relationships with the vendors.

1. Introduction

The issues associated with reaching the decision to outsource services, choosing which services you need, selecting the most appropriate vendor and working successfully with those vendors are many and varied. It is, therefore, very difficult to pull together what is a very large topic into just "relationships" between sponsor and third party laboratory. In an effort to do this, this paper will focus on some key themes which have an overall impact on the quality and success of any working relationship.

First, I will take a high level view of the current environment for sponsor companies and providers of services and what the attitude is to outsourcing amongst pharmaceutical professionals. Then, I will look more at the specifics of a working relationship, taking time to define what a good relationship actually looks like, how we go about developing and maintaining that relationship, what investments do we put in and what potential issues there can be that will cause damage.

From: *Pharmaceutical Stability Testing to Support Global Markets*: Pharmasp,
Edited by: K. Huynh-Ba, DOI 10.1007/978-1-4419-0889-6_30,

2. Current Pharmaceutical Industry Climate

It has been well documented, and is fairly obvious that the pharmaceutical industry has been going through significant and far reaching changes throughout the last 20 years. The challenges that companies faced during the 1980s were entirely different from the environment in the mid-2000s. Revenues are under pressure from high value patent expiries, lower expectations of revenues from new drugs (the apparent demise of the "blockbuster" medicine), greater regulatory scrutiny and timelines in conjunction with powerful lobbies driving down the cost of medicines to the consumers; available money is not a limitless commodity and as a result companies become more cost conscious.

To meet the changing environment, we cannot constantly increase head-counts to meet our internal demands to get molecules in to the market quicker; therefore, we need to be creative with better approaches to achieve this goal.

The global market for outsourcing continues to grow; the clinical development area is growing at a rate of about 14–15% annually with an expectation that that is going to be reaching around $24 billion by 2010. This trend is similar within the contract analytical services. Outsourcing is here to stay. To make the best use of this trend, pharmaceutical companies need to be in partnership with our service providers in order to achieve our goals.

3. Attitudes Towards Outsourcing and Contract Laboratory Services

There are a variety of attitudes within Sponsor companies towards outsourcing, at worst this can include extreme resistance to working with external service providers. It is beneficial to highlight this extreme view in order to explain the worst case scenario that could be encountered when proposing or considering outsourcing. The relevance of this to "relationships" is that if you don't tackle resistance in-house you will not realise the full potential of the relationship with the outside laboratory. There are many reasons given as arguments not to outsource a given piece of work, such as:

> "Are third party laboratories capable of doing the job to the same quality as us?"
>
> "Haven't we got the best people? Haven't we got the best knowledge and standards?"
>
> "How can we entrust this to somebody else? Don't you remember *X*?"

(Where *X* = the contract and compound of your choice where something went horribly wrong).

Steve Snyder in Contract Pharma, May 2007, following a meeting he attended regarding pre-clinical outsourcing, stated: "*Nearly every sponsor I met during the meeting shared a story of a bad outsourcing experience with me…sponsors always like to share their bad experiences*". An observable trend across all areas of outsourcing. We, as an industry, like to apportion the blame, and it would be preferable if that blame can be placed outside our own organisation.

As mentioned earlier, the level to which this view is held can vary considerably. However, if you go into a relationship with any level of negativity with the view that the contract lab could be used as a scapegoat, then at best, the performance of the select vendor is likely to be sub-optimal and below expectations or at worst, any relationship you try to build is doomed to failure.

Hopefully, if these are opinions that you have heard or even held, this may steer you away from that position. Entering a relationship from this perspective sets each party against one another from the start; it is not productive and should be challenged.

4. Defining Relationships

The key is the relationship of working together to get the same result and the best result. So what is a relationship? A dictionary definition provides a number of options (see Fig. 30-1).

The best definition for a business to business relationship is the last one, where the key word is "mutual"; shared and equal involvement. Of course, that is what a relationship is but what makes it a good one? In the simplest terms it can be broken down to trust, mutual understanding, and communication.

Firstly why bother with relationships at all, this is business not personal? Do we really care about how the other "company" feels; we are paying for it so they should do what we say? Simply, the implications of a vendor failing could be huge. If one chooses to outsource this very critical work, certainly in the registration stability stage, the potential implications are massive; delay to filing and the impact on revenue, cost of re-testing, cost of greater oversight of the lab, potentially litigation issues, etc. Therefore, there is a very real need to develop effective business relationships, *The client and vendor need to have a shared aim to get the job done within budget, to the timelines and to the highest quality.*

There are benefits to both the parties in building and maintaining good relationships. For the sponsor these include return on investment, i.e. a better level of service, greater experience and trust in vendor abilities; for the vendor these include efficiency gains as one learns to work more with a given sponsor, repeat business and increased revenues.

Going back to the definition of a relationship in this context and touching briefly on mutual understanding and communication. Understanding the business imperatives of a given company is important, and accepting that a contract laboratory is a business and needs to make money is the key factor; how one demonstrates this is something that will happen over time. In the same way, how good each party is at communicating both good and bad news will only come out through the life of a study or contract. How one deals with these things will determine how well the relationship is maintained. However, the 3rd factor in the definition "trust" and each company's attitude towards the other will define the starting point for the relationship and how well it may progress.

■ **re·la·tion·ship** (rĭ-lā'shən-shĭp') noun

- The condition or fact of being related; connection or association.
- Connection by blood or marriage; kinship.
- A romantic or sexual involvement.
- A state involving mutual dealings between people or parties or countries.

Fig. 30-1. Dictionary definition of relationship, The American Heritage Dictionary ® 4th Edition, 2000.

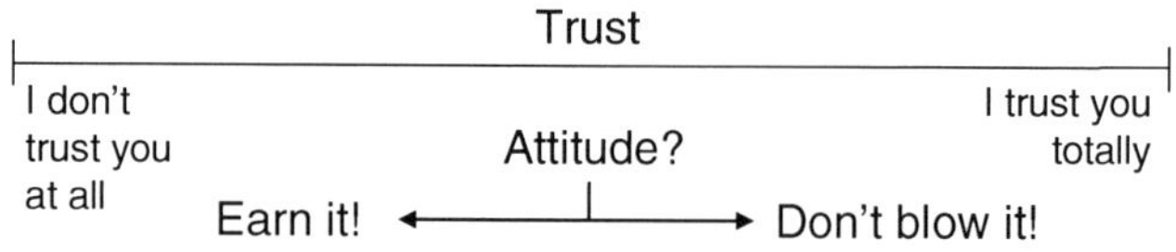

Fig. 30-2.

5. Trust

If we think very simply about trust as a continuum (as shown in Fig. 30-2), where we would put ourself at the beginning of a new business relationship when asking the question "how much do I trust this company to do the job for me?"

At each end of the continuum are the extremes, each of which guides the attitude towards the vendor and the actions when it comes to interacting with them. If the attitude is "I don't trust you at all," it could mean that "I don't trust you, earn it!" However, if the attitude is "I trust you totally", the message could well be "Don't blow it!" Either way would place the contract laboratory in a difficult position as they both have negative implications for the working relationship and management style.

The relationship with no trust would increase the chances of micromanagement, which results in increased sponsor oversight time and cost and decreases vendor ownership and motivation. In this position, the vendor is very unlikely to build up the trust; anything good that is done by them the sponsor can claim they instigated, and with decreased ownership and motivation comes a lower degree of proactivity.

If the relationship starts with complete trust, and a hands off approach, this can result in a decrease in oversight and a subsequent decrease in sponsor input. The knock on effect is that with decreased sponsor input, comes decreased vendor communication; the vendor is tacitly being given autonomy to work, which could very easily backfire if things go wrong and it needs some direction or assistance.

The aim, therefore, is the middle ground where there is a healthy respect for skills and knowledge on both sides of the relationship and realising that there will be gaps in that knowledge. The sponsor would have the most knowledge of the compound of interest, and depending on the stage that you are outsourcing, the associated methodologies. As a result expect the vendor to ask questions. A supplemental to that is contract labs should be, by their very nature, working with multiple clients and seeing different ways of working. With the right questions, the contract lab may be able to guide the sponsor on improving your own process and practises. By asking questions, one demonstrates an admirable proactive intent to avoid problems down the line.

6. Investment in the Relationship

Vendors want relationships with sponsors because it means revenue. But they will often react to the message that the sponsor gives them regarding the nature of that relationship. So the first impression counts because vendors will respond the same way mirroring the message they receive. Thus, it depends on the Sponsors to make the initial investment in making the relationship work.

As mentioned previously, money is not the only never-ending commodity. Time and knowledge are the key factors, investing time up front to plan the study or services required. What are the specific needs? What is the scope of the services and specific deliverables to be provided? At all times one should aim to ensure that there are no ambiguities and people understand what they are expected to do. However, all this needs to happen well before the vendor is contacted. Doing this pre-work internally assists in ensuring the contracting and set up processes operate more smoothly and productively.

Sharing the knowledge of the compound or study or methods shows our desire for the vendor to succeed. One furnishes all the information to do a good job, and if one is transparent, then in turn the vendor will respond with similar transparency and openness.

If a vendor is approached without such activity, the vendor can be in the dark about what is expected of them. This leads to uncertainty and potential delay. In addition, if they perform a task that was not expected or fails to perform one that was expected, then this puts strain on the relationship.

Obviously, having done the work to assess the required services, one needs to select the appropriate vendor to do the work and build the relationship. The key message regarding vendor selection is to ensure, prior to awarding the study to a specific vendor, that both parties fully understand what is expected of them. This requires a clearly defined task-specific scope with the fewest ambiguities possible.

By way of an example, I have had an experience of making a purchase of goods in a different culture that on the surface seemed very simple. However, due to a difference in communication and understanding I nearly ended up purchasing twenty times the quantity I had intended. In that example, it was easy to clarify the situation and avoid any additional costs. If this happened in a business to business relationship and it doesn't come to light quickly then this can cost a significant amount of money.

7. Maintaining the Relationship and Avoiding the Pitfalls

So, what could possibly go wrong? We assume that a project is handed over to the vendor, they will provide the services required and hand back the deliverables without any issues. Unfortunately, this may not be the case.

When thing goes wrong, one must try to rebuild the relationship. The fact is that pharmaceutical companies and sponsors love to complain and the attitude often displayed is, "one strike and you're out". We are very, very quick to judge and we do not necessarily apply the same level of behaviour in our communication to external suppliers as we would do to internal colleagues because we are paying the bills. We really need to get away from some of these ideas.

I have identified five very high level subjects that can be addressed to try and build and maintain a good working relationship. Figure 30-3 outlines a (by no means exhaustive) list of actions that can be taken to avoid any issues affecting an ongoing relationship, and build on trust, good communication and mutual understanding.

Avoiding the "Blame Culture":	• Expectations and responsibilities of each party must be clearly set from the start. • Avoid finger pointing or covering your own back – if an issue arises, deal with the problem, solve it and only afterwards see what could have been done differently to avoid problem in the future. This requires "business maturity on both sides"

Effective 2 -way communication:	• Make sure the right people talk to the right people:
	  ▪ Technical people for project issues;
	  ▪ Business people for money issues.
	• Money often gets in the way and can sour a relationship; however, it shouldn't be ignored.
	• Feedback to the vendor:
	  ▪ Relating to ongoing performance via metrics or key performance indicators; or
	  ▪ Why they have been unsuccessful at winning a bid.
	• Work to help them understand how to improve...and ask for them to do the same.
	• Listen to what the vendor is telling you.

Prepare for Problems:	• Hope for success but plan for problems.
	• The aim is to avoid major issues or minor issues escalating.
	  ▪ It costs each organisation a lot less in time, money and resources to anticipate a problem before it happens and mitigate; or if it does go awry to deal with it early.
	• Ensure structures are in place to deal efficiently with problems as and when they arise:
	  ▪ Pre-study risk assessments – Anticipate what could go wrong.
	  ▪ Issue resolution plans – Agree on contingency plans and trigger points for instigating them if things do go wrong.
	  ▪ Communication and escalation plans – Identify who is empowered to make what decisions.

Understand Each Other's Business:	• Pharmaceutical companies and Contract Laboratories, despite working in the same area are very different; however, they are both reliant on each other for success. • The issues each face are very different, as is the way in which they make their money; as a result the pressures on each vary accordingly. • Being aware of these differences and pressures can explain why some of the difficulties in a relationship occur. Demonstrating mutual understanding will build trust.

Document Agreements:	• Agree with the Contract Lab on all of the following:
	  ▪ Scope of work/Task Ownership Matrices
	  ▪ Communication, escalation, issue resolution plans
	  ▪ Deliverables
	  ▪ Costs, payment terms and schedules
	• Once agreement is reached, put them in the contract and work to them.
	• Use the contract and the agreements within it as a project management tool, not as a weapon to point out failings and not only if things go wrong and it ends in litigation!

Fig. 30-3. Key themes for building and maintaining working relationships.

Author Biography

Mr. Rob Aitchison is an Associate Director, Analytical R&D, Worldwide Pharmaceutical Sciences, based at the Sandwich site in Kent, UK. He is the Business Manager for Analytical R&D, with particular responsibilities for the outsourcing of analytical studies and tasks to third party vendors. Rob has 12 years experience in a variety of drug development and business focused roles within the Pharmaceutical industry, including contracts management and business development in the CRO arena, as well as business analysis and contracts & outsourcing within the Pharma company environment, moving to his current position in January 2006.

Chapter 31

Outsourcing Stability Testing: A Tool for Resource and Risk Management

Michael D. Barron

Abstract A decision model that has been developed to use when evaluating outsourcing stability studies as resource management is a key driver. However, International Conference of Harmonization (ICH) Q9 provides insights into applying risk management principles into the process with noteworthy results. This paper explores many of the practical considerations in the decision-making process and the advantages and disadvantages of which companies have used when making a decision to outsource. These considerations will be re-framed in the model of risk-management as discussed in ICH Q9.

1. Introduction

This article is derived from a brief presentation designed to stimulate discussion in an interactive workshop. It is not an exhaustive discussion of the topic, but does cover some important aspects of outsourcing stability testing. It was presented as a collaborative adjunct to the views expressed in the other two articles derived from the AAPS interactive workshop on outsourcing stability studies and is intended to be predominantly from the Contract Research Organization (CRO) point of view. This includes a brief discussion of operating, regulatory, and business environments in which a product is tested throughout its lifecycle. It also briefly addresses some key factors affecting successful stability outsourcing and risk assessment according to the ICH Q9 document.

2. Operating Environments

2.1. Regulatory Environment

Stability studies are required during every phase of a drug's product lifecycle. They are integral for drug development decision making, clinical use, expiration dating, and regulatory approval. Stability studies also serve as a quality control function throughout the commercial life of the product.

From: *Pharmaceutical Stability Testing to Support Global Markets*: Pharmasp,
Edited by: K. Huynh-Ba, DOI 10.1007/978-1-4419-0889-6_31,

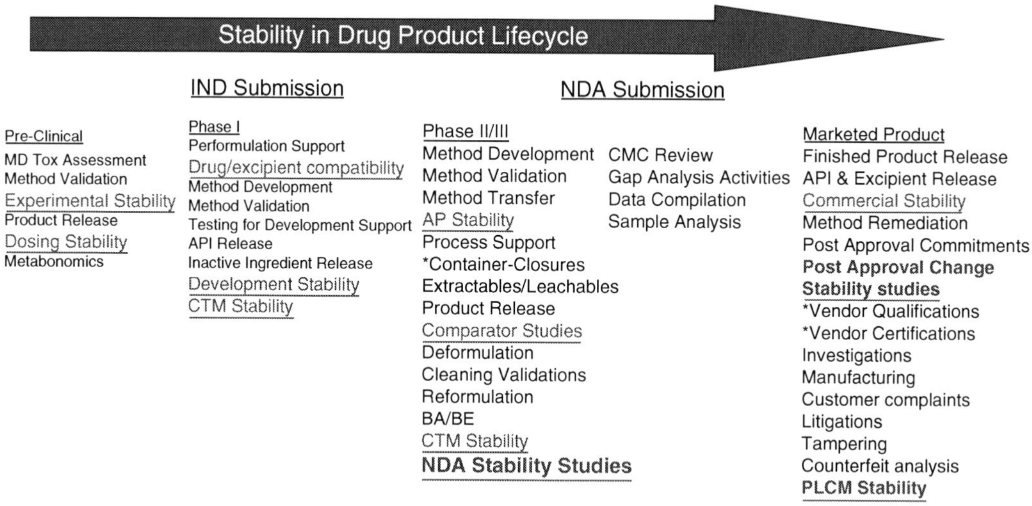

Fig. 31-1. Stability in drug product lifecycle

Stability studies in pre-Investigational New Drug (pre-IND) and early clinical phases are not high resource demands in comparison to the stability studies required for product registration and marketing approval. ICH compliant registration stability protocols demand significant resources for storage and testing facilities as they typically require three lots of each strength and package configuration in multiple storage environments. This large resource demand can strain the capacity of in-house labs that often have a drug pipeline support role.

Commercial product and then Product Life Cycle Management (PLCM) stability studies come after the approval of the product. PLCM covers new dosage forms, modified releases, new packaging, etc. (see Fig. 31-1).

The ICH Q9: Quality Risk Management Guideline was published in the Federal Register, Vol. 71, No 106, pages 32,105–32,106, June 2, 2006. This guideline provides principles and examples of tools for quality risk management that can be applied to all aspects of pharmaceutical quality including development, manufacturing, submission/review processes, distribution, and inspection throughout the lifecycle of drug substances and drug products, including the use of raw materials, solvents, excipients, packaging and labeling materials.

2.2. Business Environment

The Contract Research Organization (CRO) conducts stability studies as a profit center where typical arrangements are fee-for-service or full time equivalent (FTE) programs. Very simply stated, one can look at the revenue being based on dollars per revenue generating headcount and prices are based on fully burdened costs plus profit margin. Managing contract lab capacity to meet demand is an on-going business challenge.

Pharmaceutical companies (Pharma), including large, medium, small, virtual, biotech, specialty, generic, consumer health, and nutritional companies conduct stability studies as a cost center. Revenue is primarily based on product sales, development milestone payments, or in-licensing valuations and the costs are not necessarily allocated in the same manner as a CRO. Therefore, fully burdened costs are sometimes difficult to know.

Both parties have a goal to manage a profitable business. For CRO's , the main objective, simply stated, is to manage profit (revenues – fully burdened costs) to a level equal to or greater than the potential returns on alternative investments, while remaining competitive.

The value proposition of outsourcing for Pharma, (benefit of outsourcing - cost of outsourcing) should be greater than the cost of keeping the work in-house. In many situations, a significant benefit is realized as cost (or risk) avoidance. An example of a significant business risk to Pharma management is sustaining labor, laboratory, and stability storage infrastructure costs to meet peak demands on stability related resources. Those same costs are then carried through the valleys of low resource demand.

Personnel, equipment, and facilities are fixed costs if owned. There are financial benefits gained by optimizing the mix of fixed and variable costs. Outsourced work can be accounted for as variable cost. This is favorable to rapid capacity scale-up and capacity ramp-down without incurring long term fixed costs, depreciation, and related headcount issues.

3. Key Factors

There are essential elements to all stability projects whether outsourced or not; these include the purpose of the study, the test methods and specifications (which are specific to the dosage form, delivery system, and packaging), storage conditions, time-points, sample quantity requirements, data reporting and statistical evaluation when appropriate. A good quality CRO can handle all aspects of a stability study.

A CRO considers whether the study sponsor is an existing or new customer. Existing confidentiality, quality, and master service agreements, or other standard terms and conditions facilitate new study initiation. Accepting a new customer incurs thousands of dollars of costs to the CRO and Pharma alike to initiate that new relationship, e.g., interactions with company lawyers to get agreements in place plus scheduling and conducting a quality audit. There are also business and technical team discussions among the many interactions to take place before the work of the lab begins.

The degree of attractiveness of the project to the CRO is in the scope of the request for proposal (RFP); i.e., it is a one-off, short-term study, full method development and validation plus stability, part of an integrated product development project, or commercial product support, etc

The extent of the existing relationship and potential for the new relationship are important considerations. Extensive involvement with multiple studies often requires a Master Service Agreement and full Quality Agreement. A CRO should also consider the available capacity and if the scope of work is a good fit; that includes timing, duration, personnel, capability, equipment, facilities, etc.

It is important to the relationship to differentiate between those who influence decisions and those who make decisions, e.g., authority to sign contracts and test protocols. Sometimes there are gaps between the perceived benefits for Pharma scientific operations and their supporting procurement department. It is not difficult to understand that Pharma scientific operations may place different levels of value to the priorities of time, quality, scientific expertise, and existing relationships versus the competitive price. By design, procurement has competitive pricing as a priority.

When it comes to time as a risk factor, consider the various aspects of time. These include the duration of the project; either through Turn Around Time (TAT) for certain activities and the corresponding need for On Time Delivery (OTD) for priority timepoints, e.g., clinical start dates and filing deadlines. Another important aspect of time and part of project success is the timeliness of reporting problems as they may arise. This influences the ability of Pharma to be part of the solution and minimize impact to other parts of their organization. Other considerations include the time invested into managing the ongoing relationship plus the potential for accelerating the product development timeline (or just enabling an aggressive timeline dictated by Pharma management).

For quality assurance, consider the compliance history and robustness of compliance systems; education, training and experience of staff; and the degree of flexibility to meet your unique requirements. Keep in mind however that the more exceptions to standard CRO practices Pharma makes with the CRO, the greater the risk of non-conformance to those special standards. It is vitally important to have open and frequent communication to build the trust needed for successful relationships.

Finally, consider the long term stability of the CRO itself. If you are placing a 3–5 year stability study with a CRO, you have a vested interested in their long-term viability as a business.

4. Risk Assessment

In making outsourcing decisions, it may be a good idea to construct a risk matrix utilizing the risk ranking and filtering tool as defined in the ICH Q9 document. It's a good guidance document and is recommended reading (see Figs. 31-2 and 31-3 for examples of rank matrices taken from ICH Q9).

To construct such a matrix, it is recommended to involve cross functional members of the team because of the different ways people view the activities and assess the risks. It is mandatory to consider the different perspectives and priorities of the quality assurance person, the lab manager, the project manager, the finance person, the manufacturing site, and the clinical researchers.

Building the matrix requires identification of what might go wrong based on information, experience and opinion. After constructing the matrix, the analysis is conducted. Estimate the risk based upon likelihood of occurrence and severity of impact if it does happen. Define risk classifications and how

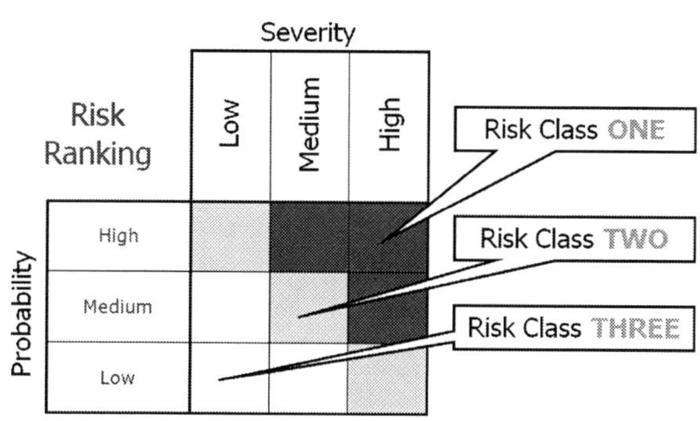

Fig. 31-2. Probability vs. severity ranking matrix (ICH Q9)

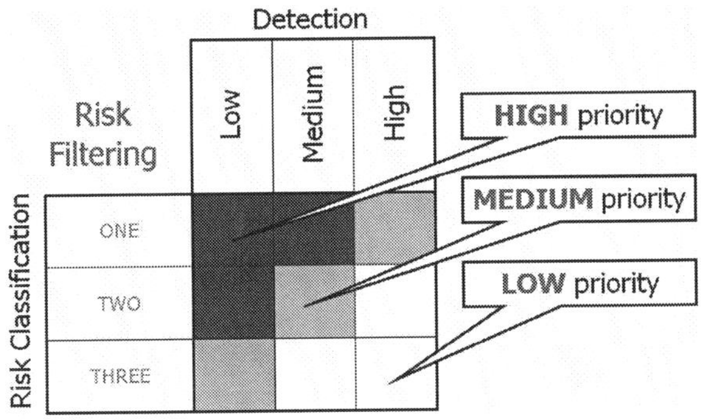

Fig. 31-3. Risk classification vs. detection ranking matrix (ICH Q9)

detectable those risks are. Implement controls, communication plans, and periodic reviews of the effectiveness of the program.

5. Conclusion

Outsourcing stability testing decisions are affected by the operating environments in which the product is being produced. These are primarily the regulatory and business environments. In the United States, the regulatory environment is bounded by IND, NDA, ANDA, GMP etc., plus PLCM strategies involving stability tests throughout the product's lifecycle. The business environment is driven by the relationship between the CRO and the customer, as well as the optimization of the value proposition which includes benefits, revenues, costs, and risks of doing business together.

Utilizing a cross functional team to build a ranking matrix for risk assessment can be effective to capture all aspects of risk due to the different ways people evaluate the risk classifications according to their specialty and role in the development and manufacturing process.

Other key factors affecting the outsourcing process are the multiple aspects of time and quality combined with other aspects of successful relationships driven by frequent, open, effective communication.

Author Biography

Mr. Michael Barron has 20 years experience in contract pharmaceutical development and testing. He was director of stability services for Pharmaceutical Technologies and Services at Cardinal Health, Research Triangle Park, NC. He is a Kaizen Leader within the Lean Six Sigma context. He has presented internationally on Operational Excellence in Stability Operations and Protocol Design for International Product Registrations. His responsibilities included all aspects of multi-site stability sample storage operations, R&D and commercial product protocol development and implementation, and consulting on stability programs. He sometimes serves as a facilitator for national meetings of the Pharmaceutical Stability Discussion Group (PSDG), has developed and moderated numerous accredited seminars on drug development, and published a chapter on implementing successful stability testing operations in the Analytical Chemistry and Testing Primer for Pharmaceutical Technology.

Index